Applications of Blockchain Technology

This book explores the transformative potential of blockchain technology and shows how blockchain can promote equitable access to resources and bridge societal gaps. It examines blockchain's role in enhancing transparency, security, and decentralization across various sectors and in society. The book addresses both the advancements and challenges of blockchain, providing practical use cases to understand the opportunities, limitations, and innovative applications of blockchain, envisioning a future where this technology drives significant societal change.

- Showcases practical use cases in healthcare, finance, governance, supply chain management, and humanitarian aid
- Covers decentralized finance (DeFi), financial inclusion, asset tokenization, and impact investing
- Investigates blockchain's role in environmental conservation, charitable donations, and ethical considerations
- Examines scalability, interoperability, security, and regulatory concerns and discusses self-sovereign identity and personal information control through blockchain
- Offers insights for policymakers on creating a conducive environment for blockchain while safeguarding public interests

This is a reference for scholars and practitioners interested in the impact of blockchain.

Applications of Blockchain Technology

Edited by

Arvind Mewada, Nagendra Singh,
Mohd. Aquib Ansari, and Amrendra Singh Yadav

CRC Press
Taylor & Francis Group
Boca Raton London New York

CRC Press is an imprint of the
Taylor & Francis Group, an **informa** business
A CHAPMAN & HALL BOOK

Designed cover image: ShutterStock

First edition published 2026
by CRC Press
2385 NW Executive Center Drive, Suite 320, Boca Raton FL 33431

and by CRC Press
4 Park Square, Milton Park, Abingdon, Oxon, OX14 4RN

CRC Press is an imprint of Taylor & Francis Group, LLC

ISBN: 9781032899862 (hbk)
ISBN: 9781032899879 (pbk)
ISBN: 9781003545620 (ebk)

DOI: 10.1201/9781003545620

Typeset in Times
by codeMantra

Contents

SECTION 3 Revolutionizing Finance

SECTION 4 Societal and Industrial Impacts

Editors

Dr. Arvind Mewada is an Assistant Professor at Bennett University, Greater Noida, India. He earned his doctorate from the Motilal Nehru National Institute of Technology, Allahabad, Uttar Pradesh, India. He holds an M.Tech. in Computer Science and Engineering from the Maulana Azad National Institute of Technology, Bhopal, and a B.E. in Information Technology from Rajiv Gandhi Proudyogiki Vishwavidyalaya, Bhopal, India. His research focuses on fake review detection, reviewer analysis, and group detection based on natural language processing, machine learning, and deep learning. His research interests include natural language processing, deep learning, distributed computing, blockchain technology, and ledger technology. He has published more than 25 research articles in reputed journals (SCI/SCIE indexed) and conferences. Dr. Mewada has been an expert speaker in various faculty development programmes and has organized numerous conferences, workshops, and short-term courses.

Google Scholar ID link: https://scholar.google.co.in/citations?user=226MeMcAAAAJ&hl=en

Dr. Nagendra Singh is the Principal of the Trinity College of Engineering and Technology, Karimnagar, Telangana, India. He has worked in various positions, such as Head of Department (HOD), Dean of Academics, and Exam Superintendent. He has published many research articles in reputed refereed SCI and Scopus-indexed journals and conferences. He has published 11 book chapters and edited five books published by Indian and international publishers. Out of five books, two books were edited for CRC Press. He has been granted 14 design patents. He has organized many national and international conferences as well as faculty development programmes and seminars. Under his guidance, one candidate completed his Ph.D., and three candidates completed their M.Tech. in Electrical Engineering. His areas of interest are optimization techniques, AI, power systems, renewable energy systems, power electronics, smart antennas, and blockchain technology. He is a life member of various professional bodies. He is a reviewer for many international journals.

Google Scholar Id link: https://scholar.google.com/citations?user=xd7O-GIAAAAJ&hl=en

Dr. Mohd. Aquib Ansari earned a B.E. degree in Information Technology from SATI, Vidisha, in 2014. He completed his M.Tech. in Information Technology at MITS, Gwalior, in 2017. He earned his Ph.D. degree from MNNIT Allahabad, Prayagraj, India, in 2022. Currently, he is an Assistant Professor at SCSET, Bennett University, Greater Noida, India. He has published more than 28 research papers in various international journals and conferences. His research interests include computer vision and image processing, block chain technologies, artificial intelligence, and machine learning.

Google Scholar ID link: https://scholar.google.com/citations?user=IMWJqrAAAAAJ&hl=en

Dr. Amrendra Singh Yadav is an Assistant Professor at Atal Bihari Vajpayee Indian Institute of Information Technology and Management, Gwalior, India. He earned his doctorate from the Motilal Nehru National Institute of Technology Allahabad, Uttar Pradesh, M.Tech. from the Madan Mohan Malaviya University of Technology, Gorakhpur, Uttar Pradesh, and B.Tech. in Information Technology from State University, Uttar Pradesh, India. His research focuses on improving the scalability of blockchain networks, developing use-case-specific consensus algorithms, and the security of blockchains. His research interests include blockchain technology, peer-to-peer systems, consensus algorithms, distributed ledger technology, distributed systems, cloud computing, and computer networks. He has published more than 30 research articles in reputed journals (SCI/SCIE) and indexed journals and conferences. Dr. Yadav has been an expert speaker in various faculty development programmes and has organized a number of conferences, workshops, and short-term courses.

Google Scholar ID link: https://scholar.google.com/citations?hl=en&user=vHFpDxoAAAAJ

Contributors

Shyam Sunder Agrawal
Mittal School of Business
Lovely Professional University
Phagwara, India

Shahnawaz Ahmad
School of Computer Science
Engineering and Technology
Bennett University
Greater Noida, India

Saaema Akhtar
Department of Human Resources
Vivo Middle East headquarters
Dubai, United Arab Emirates

Annu
Department of Humanities and Social Sciences
Motilal Nehru National Institute of Technology Allahabad
Prayagraj, India

Mohd. Aquib Ansari
School of Computer Science
Engineering and Technology
Bennett University,
Greater Noida, India

Sumit Badotra
Computer Science and Engineering
National University of Singapore,
Singapore

N. Sharmila Banu
School of CS and AI
SR University
Warangal, India

Bhawna
Department of Information Technology
Delhi Technological University,
Delhi, India

Mohd Bilal
Department of Electrical Engineering
Aligarh Muslim University,
Aligarh, India

Mohd Sameen Chishti
Department of Computer Science
Norwegian University of Science and Technology (NTNU),
Trondheim, Norway

K. Deepa
School of CS and AI
SR University,
Warangal, India

G Uma Devi
Computer Science and Engineering
University of Engineering and Management
Jaipur, India

Zenab Elgamal
School of Computing
Skyline University College, University City of Sharjah
Sharjah, United Arab Emirates

Khalid Abdullah Fakeeh
Department of Information Systems
King Abdulaziz University, KAU
Jeddah, Saudi Arabia

Prafull Farswal
School of Computing Science Engineering and Artificial Intelligence
VIT Bhopal University,
Kothri Kalan, India

Salman Hameed
Department of Electrical Engineering
Aligarh Muslim University
Aligarh, India

Shahadat Hussain
School of Computer Science Engineering and Technology
Bennett University
Greater Noida, India

Imdadullah
Department of Electrical Engineering
Aligarh Muslim University
Aligarh, India

Garima Jain
School of Computing Science Engineering and Artificial Intelligence
VIT Bhopal University,
Kothari Kalan, India

Sanat Jain
School of Computing Science Engineering and Artificial Intelligence
VIT Bhopal University,
Kothari Kalan, India

Niharika Keshari
School of Computer Science Engineering and Technology
Bennett University,
Greater Noida, India

Asad Khan
School of Computer Science Engineering and Technology
Bennett University,
Greater Noida, India

Pravin R Kshirsagar
Department of Electronics and Telecommunication Engineering
J D College of Engineering and Management,
Nagpur, India

J. Sathish Kumar
Department of Computer Science and Engineering
Motilal Nehru National Institute of Technology Allahabad,
Prayagraj, India

Sushil Kumar
Department of Humanities and Management
Dr. B R Ambedkar National Institute of Technology Jalandhar,
Jalandhar, India

Vishal Kumar
Department of Humanities and Management
Dr. B R Ambedkar National Institute of Technology Jalandhar
Jalandhar, India

M L S N S Lakshmi
Electronics and Communication Engineering
Ramachandra College of Engineering
Eluru, India

K. Lalitha
Department of Electronics and Communication Engineering
SRM Institute of Science and Technology
Chennai, India

K. M. Manisha
Department of Computer Science and Engineering
Motilal Nehru National Institute of Technology Allahabad,
Prayagraj, India

Balajee Maram
Department of Computer Science and Artificial Intelligence
SR University,
Warangal, India

V. Megala
Department of Electronics and Communication Engineering
SRM Institute of Science and Technology
Chennai, India

Arvind Mewada
School of Computer Science Engineering and Technology
Bennett University,
Greater Noida, India

Divya Mobarsa
Department of Mathematics
Marwadi University
Rajkot, India

Jiyaul Mustafa
Department of Mechanical Engineering
Bennett University,
Greater Noida, India

Yuri Nikolaichik
Foundry Machinery and Technology
Mechanical Technological Faculty
Belarusian National Technical University
Minsk, Belarus

Satyam Omar
SCSET
Bennett University,
Greater Noida, India

Nimmagadda Chandra Paul
Department of Electronics and Communication Engineering
Ramachandra College of Engineering,
Eluru, India

Greetta Pinheiro
School of Computer Science Engineering and Technology
Bennett University,
Greater Noida, India

Manvi Prajapati
Department of Mathematics
Axis Colleges
Kanpur, India

J. S. Prasath
Department of Computer Science and Engineering
Sapthagiri NPS University
Bengaluru, India

Saloni Raheja
Mittal School of Business
Lovely Professional University
Phagwara, India

Harsh Raj
School of Computing Science Engineering and Artificial Intelligence
VIT Bhopal University
Kothari Kalan, India

B Gangadhara Rao
Department of Electronics and Communication Engineering
Ramachandra College of Engineering,
Eluru, India

Swati Rawal
Research and Development Department
Ernst and Young
London, United Kingdom

Kalpana Rawat
School of Business Management
Noida International University,
Noida, India

K. Sankar
Computer Science and Engineering
CVR College of Engineering
Mangalpalle, India

Amit Sata
Department of Mechanical Engineering
Marwadi University
Rajkot, India

Rajasekaran Selvaraju
Department of Information Technology
University of Technology and Applied Sciences,
Muscat, Oman

Sana Shahab
Department of Business Administration, College of Business Administration
Princess Nourah bint Abdulrahman University
Riyadh, Saudi Arabia

Mastan Vali Shaik
Department of Information Technology
University of Technology and Applied Sciences,
Muscat, Oman

Shaik Saidhbi
Computer Science and Engineering
Samara University
Semera, Ethiopia

Sumit Kumar Sharma
Department of Information Technology
Delhi Technological University
Delhi, India

Minal Shukla
Research and Development Department
MGL Group
Rajkot, India

Vinay Shukla
Department of Mathematics, SCSET
Bennett University,
Greater Noida, India

Kedar Nath Singh
Department of Computer Science and Engineering
Amity University
Gwalior, India

Nagendra Singh
Department of Electrical and Electronics Engineering
Trinity College of Engineering and Technology,
Karimnagar, India

Reetu Singh
School of Computer Science Engineering and Technology
Bennett University,
Greater Noida, India

Vartika Singh
School of Computing Science Engineering and Artificial Intelligence
VIT Bhopal University
Kothari Kalan, India

Richa Srivastava
School of Business Management
Noida International University,
Noida, India

Amanjot Singh Syan
Mittal School of Business
Canadore Stanford International College
North Bay, Ontario, Canada

Namita Tiwari
Department of Mathematics, School of Basic Sciences
CSJM University
Kanpur, India

Ravindra Tripathi
Department of Humanities and Social Sciences
Motilal Nehru National Institute of Technology Allahabad
Prayagraj, India

Arvind Kumar Upadhyay
Department of Computer Science and Engineering
Amity University
Gwalior, India

Richa Verma
School of Business Management
Noida International University,
Noida, India

Utkarsh Kumar Verma
School of Business Management
Noida International University,
Noida, India

Ahsan Waseem
Department of Electrical Engineering
Aligarh Muslim University
Aligarh, India

Mohammed Wasid
School of Computer Science Engineering and Technology
Bennett University,
Greater Noida, India

Amrendra Singh Yadav
Department of Computer Science and Engineering
Atal Bihari Vajpayee Indian Institute of Information Technology and Management
Gwalior, India

General Introduction

Blockchain technology, introduced in 2008 as the foundation for Bitcoin, has since emerged as a transformative innovation with applications extending far beyond cryptocurrency. It provides a decentralized, transparent, and tamper-proof ledger system that addresses critical issues like data integrity, trust, and security in digital interactions. Blockchain's potential to eliminate intermediaries and enable peer-to-peer transactions has implications for sectors such as finance, healthcare, supply chain management, and governance. For example, blockchain can enhance transparency in supply chains by enabling real-time tracking of goods, ensuring authenticity and ethical sourcing. Similarly, it can revolutionize governance systems through secure and transparent voting mechanisms, fostering trust in democratic processes.

Despite its promise, blockchain faces significant challenges that limit its broader adoption. Scalability remains a pressing concern, with many networks struggling to match the transaction speeds of traditional systems. In addition, issues like high energy consumption, regulatory uncertainty, and difficulties in integrating blockchain with existing infrastructures pose barriers.

This book aims to explore the dual nature of blockchain's societal impact—its ability to drive progress while contending with persistent challenges. By critically examining its applications, limitations, and potential, this study seeks to provide actionable insights to guide stakeholders in maximizing blockchain's benefits while addressing its risks. The findings will contribute to shaping policies and strategies that ensure blockchain becomes a powerful tool for societal advancement, promoting transparency, efficiency, and inclusivity across various domains.

This book offers readers the opportunity to engage with cutting-edge technologies in the blockchain. It consists of a collection of 20 chapters, grouped into five key themes:

- Healthcare Transformation
- Integration with Emerging Technologies
- Revolutionizing Finance
- Societal and Industrial Impacts
- Security and Cryptography

This structure provides a comprehensive overview of the current state and future directions of block technologies in the context of society development.

SECTION 1: HEALTHCARE TRANSFORMATION

This section deals with possible healthcare transformation in blockchain. Blockchain's role in the healthcare sector is revolutionary, as highlighted in several chapters. The technology enables secure data sharing, improves medical record analytics, and streamlines healthcare data management. These chapters emphasize blockchain's ability to enhance patient care by addressing privacy concerns and introducing secure frameworks for maintaining and sharing sensitive patient information. Specific use cases include blockchain's application in drug traceability, ensuring data integrity in clinical trials, and enabling interoperability among healthcare providers. The integration of blockchain with the Internet of Things (IoT) in healthcare is also discussed, demonstrating how intelligent systems can further optimize patient care and operational efficiency.

SECTION 2: INTEGRATION WITH EMERGING TECHNOLOGIES

This section includes topics related to the integration of blockchain with emerging technologies. Blockchain's synergy with emerging technologies like the IoT and artificial intelligence (AI) is a recurring theme. Chapters in this section explore the architectural integration of blockchain with IoT to address challenges such as data scalability, security, and privacy. Intelligent learning models powered by AI are examined for their potential to enhance blockchain-enabled systems. These chapters also delve into how the convergence of these technologies facilitates efficient data analysis, secure decision-making, and the development of self-learning networks. The integration of blockchain and IoT is presented as a solution to long-standing challenges, creating systems that are secure, scalable, and efficient.

SECTION 3: REVOLUTIONIZING FINANCE

In this section, blockchain concepts in finance are explored. The financial sector has been profoundly impacted by blockchain technology, as extensively analysed in this volume. The chapters in this section provide insights into blockchain's disruptive influence on traditional financial frameworks. From streamlining international payments to enabling decentralized finance platforms (DeFi), blockchain is transforming financial operations. The chapters discuss how blockchain enhances transparency, reduces costs, and increases efficiency in financial transactions. In addition, the potential of blockchain to enable asset tokenization, smart contracts, and decentralized autonomous organizations is explored, showcasing its ability to modernize and democratize the financial sector. Regulatory challenges, scalability issues, and future research directions are also addressed, providing a balanced view of the opportunities and limitations in this domain.

SECTION 4: SOCIETAL AND INDUSTRIAL IMPACTS

In this section, the societal and industrial impacts of blockchain are explored. Three chapters are presented in this section. Blockchain's applications extend beyond traditional sectors to address societal and industrial challenges. This section highlights blockchain's role in fostering sustainable energy trading within microgrids, improving supply chain transparency, and revolutionizing biotechnology. Chapters examine the transformative impact of blockchain on social governance, emphasizing its potential to enhance accountability and equity in management practices. Use cases include blockchain's application in seed supply chain management, which improves traceability and reduces inefficiencies, and its role in supporting corporate social responsibility and social impact investing. The volume underscores blockchain's ability to drive sustainability initiatives and create ethical, transparent systems across industries.

SECTION 5: SECURITY AND CRYPTOGRAPHY

This section briefly explores the security concerns and cryptography used in blockchain. Security is a critical aspect of blockchain technology, and several chapters are dedicated to exploring its mechanisms and challenges. These chapters provide a detailed analysis of how blockchain ensures resilience against cyber threats through robust cryptographic techniques. Topics include the use of hash functions, digital signatures, and public key cryptography to maintain data integrity and trust fewer consensuses among network participants. The chapters also address challenges such as scalability, energy consumption, and the regulatory landscape, proposing solutions to enhance the security and efficiency of blockchain systems. Real-world applications, such as secure supply chain management and smart contract deployment, are discussed to illustrate the practical implications of blockchain's security features.

Section 1

Healthcare Transformation

1 Harnessing Blockchain Technology for Enhanced Healthcare Data Management and Analytics

K. Lalitha, V. Megala, and Khalid Abdullah Fakeeh

1.1 INTRODUCTION

Timely interchange of patient data among diverse healthcare services significantly reduces costs, improves treatment, and strengthens disease surveillance [1]. Transmitting medical images or patient data through a centralised data centre is inappropriate, compromising storage and security settings [2]. The lack of secure connections to integrate several independent health systems into an accessible network while protecting the privacy of healthcare professionals constitutes a significant healthcare challenge. The Ethereum blockchain platform was established to demonstrate the efficacy of blockchain in managing sensitive data across various sectors, including finance, information technology, healthcare, security, real estate, the Internet of Things, and public services [3]. Blockchain offers solutions to critical healthcare issues, including secure medical record interchange and adherence to data protection regulations. Blockchain is a technology developed in 2008 as part of the Bitcoin proposal to alter or, more precisely, supplant electronic transactions [4,5]. Blockchain transactions are generated and preserved chronologically; participants can monitor digital money and user information without relying on centralised record-keeping. This technology's key attribute is the capability to store a complete copy of the blockchain on the nodes participating in the network [6].

Moreover, each transaction is traceable, hence complicating the perpetration of fraud. The term "blockchain" lacks a broadly accepted definition due to its complex nature, which includes technological, operational, legal, and regulatory aspects [5,7]. Blockchain is a technology developed in 2008 as a component of the Bitcoin proposal to alter or, more precisely, supplant electronic transactions [8]. Due to the historical creation and maintenance of blockchain transactions, users can monitor digital currency and patient information without the need for centralised record-keeping [9]. The fundamental attribute of blockchain is its ability to store a complete copy of the entire blockchain on the nodes participating in the network [10].

DOI: 10.1201/9781003545620-2

Moreover, each transaction is traceable, hence complicating the perpetration of fraud. It is crucial to highlight that "blockchain" lacks a universally agreed definition because of its multifaceted nature, encompassing technological, operational, legal, and regulatory dimensions. The implementation of blockchain technology might significantly revolutionise the healthcare sector through crucial opportunities such as decentralised storage, which allows data to be transparent and accessible to third parties with the creator's consent. Maintaining many copies of data across diverse places constitutes a decentralised approach to data storage [11]. Consent is the global consensus algorithm that will govern access, storage, and distribution. Data modification means that data alterations may only occur with the autonomous permission of all relevant parties. Immutability means that the modification of any patient data is prohibited. Data stored in a certain block is immutable and cannot be modified or amended at any time. Enhanced capacity means upholding data privacy using blockchain, which is cost-effective due to the absence of intermediaries and the streamlined process for obtaining permissions [12].

Section 1.2 addresses the rationale for adopting blockchain technology in healthcare, while Section 1.3 outlines the fundamental aspects of the blockchain deployment process. Section 1.4 addresses potential blockchain implementation strategies for healthcare sub-domains. Section 1.5 examines the issues associated with blockchain technology.

The main objective of this study is to investigate and analyse blockchain technology in connection to healthcare applications. This study also investigates the requirements of blockchain and the main difficulties in building blockchain-integrated healthcare systems. In addition, the importance of smart contracts in healthcare systems is explored. Blockchain is a network of authenticated users or nodes that manages a decentralised public ledger database. It has immutable data blocks that can be securely transferred without the help of third-party signatures and consensus in cryptography.

1.2 BLOCKCHAIN TECHNOLOGY IN HEALTHCARE

1.2.1 Distributed Ledger Technology

Blockchain, a form of distributed ledger technology (DLT), is illustrated in Figure 1.1. The fundamental concept of DLT is that all users have access to a shared database. Bitcoin introduced the notion of an open, universally accessible ledger, providing the initial solution to the difficulty of establishing confidence in an unsafe environment without reliance on a third party. DLT designates stockholders as nodes possessing two keys, with a hash generated by a cryptographic method that interconnects all blocks on the blockchain (e.g., SHA256). It guarantees that the block is anonymous, immutable, and concise. This employs a private key to authenticate a transaction digitally and guarantees its legitimacy. The first stipulation is that the transaction can solely be authenticated with a certain key, while the second condition is that an error during the transfer inhibits decoding.

The consensus process regulates the data incorporated in the block and the selection of miners. A peer-to-peer system is established, wherein each node is designated

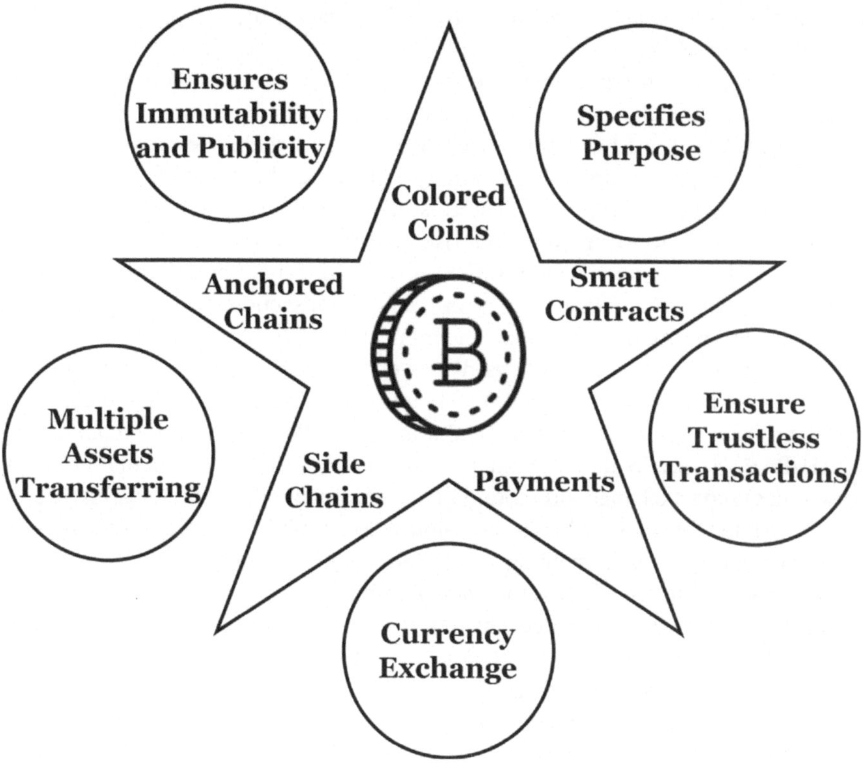

FIGURE 1.1 Distributed ledger technology.

as a peer upon initiating communication and interaction with other nodes in the network; thus, it is referred to as a peer node. A full node in the network refers to any device that possesses the whole blockchain host state and is accountable for operating the entire blockchain system. When a client intends to utilise the blockchain, it connects to the network through a node. Blockchain technology in healthcare has numerous critical components that collectively facilitate its uses in securing, managing, and exchanging healthcare data.

1.2.2 Essential Components of Blockchain for Healthcare

The fundamental elements of blockchain in medical applications highlight key components and their specific functionalities. The decentralised ledger is the core of blockchain, storing transactions across a network securely and transparently. In healthcare, it encrypts and preserves patient records, treatment histories, and other sensitive data, with each transaction being timestamped, cryptographically linked, and distributed throughout the network. Cryptographic hashes transform data into unique, fixed-size strings, helping to identify and maintain data integrity within the blockchain. Any data alteration would result in a new hash, alerting the network to potential tampering. Smart contracts are self-executing agreements with predefined rules, useful for managing patient consent, insurance processing, and regulatory compliance in healthcare.

They facilitate transparency and enable trustless transactions without intermediaries. Consensus mechanisms, such as Proof of Work (PoW), Proof of Stake (PoS), and Practical Byzantine Fault Tolerance (PBFT), ensure that nodes across the blockchain network agree on transaction validity, enhancing the reliability and accuracy of medical records. Encryption transforms plain text into ciphertext, safeguarding healthcare data by allowing only authorised parties to access and decrypt sensitive information. The immutable audit trail ensures that it cannot be altered once data is verified. It creates an auditable history of medical transactions that strengthens data integrity, reduces fraud, and supports regulatory compliance. Permissioned access allows controlled entry to blockchain networks, ensuring that sensitive medical information remains secure. Patients can manage who has access to their data, aligning with privacy regulations such as Health Insurance Portability and Accountability Act (HIPAA). Interoperability standards, such as HL7 and FHIR, define protocols for seamless data exchange, promoting blockchain integration with existing healthcare IT systems for improved data accessibility and usability. Addressing scalability and performance is vital, enabling blockchain solutions to handle large-scale data transactions without sacrificing efficiency or security.

Lastly, regulatory compliance ensures that blockchain applications in healthcare adhere to data protection regulations, preserve patient rights and privacy, and meet legal obligations for healthcare providers. Together, these components provide a secure, transparent, and efficient framework for blockchain applications in the medical field. Figure 1.2 delineates the fundamental functions of a blockchain node. These essential functions ensure that blockchain networks are decentralised, secure, and reliable. The blockchain node, characterised by decentralisation, enhanced security, and transparency, is revolutionising finance, supply chain management, and

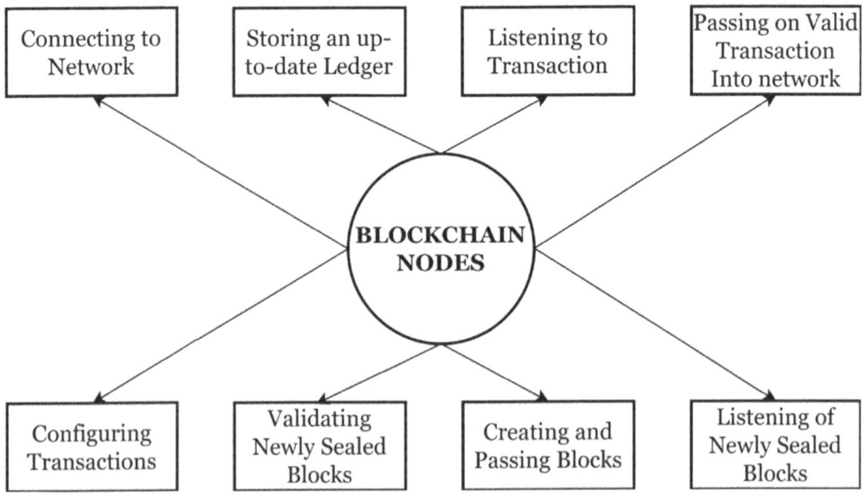

FIGURE 1.2 Types of Block chain nodes.

healthcare sectors. Its ability to eradicate intermediaries, ensure data integrity, and provide tamper-proof records drives innovation across all sectors, offering more efficient, secure, and transparent solutions to traditional challenges. This transformative impact has established blockchain as a cornerstone for future developments across multiple domains [13,14].

1.2.3 Blockchain Classification

Blockchain is categorised into four main types: public, private, hybrid, and consortium. A public blockchain enables all participants to access comprehensive information and engage in the consensus mechanism. A private blockchain is confined to a designated group. In a private blockchain, all nodes within a specific group are authorised to engage in the consensus process [15]. A hybrid blockchain environment amalgamates the advantages of both public and private blockchain networks. Consortium blockchains are managed by a collective of organisations rather than an individual entity. They provide shared control yet remain permissioned, indicating that only designated individuals can access the blockchain. This illustrates how the public blockchain integrates the security and confidentiality of private blockchains with the transparency of blockchain networks [16,17]. In blockchain applications, we must resolve two challenges: double spending and the Byzantine Generals Problem. The double spending issue arises when a single currency is used in two concurrent transactions. Blockchain addresses this issue by enabling numerous decentralised nodes to authenticate transactions. This necessitates the development of a corresponding consensus algorithm.

1.2.4 Consensus Algorithms

- **Proof-of-Work:** It is mostly linked to blockchain technology used in Bitcoin. Miners vie to address a computationally demanding task while employing the Proof-of-Work (PoW) protocol. The miner that assesses this hash value is compensated for initially validating the transactions or other elements within the block. The PoW protocol necessitates substantial energy consumption when implemented on a large blockchain.
- **Proof-of-Stake:** The objective of developing Proof-of-Stake (PoS) is to address the issue of energy inefficiency. The fundamental premise is that nodes possessing a larger stake will exhibit an increased likelihood of contributing blocks to the network. A new block signer is selected randomly, with the selection procedure determined by the miner's stake amount. As a result, the energy required to mine a block will be markedly reduced relative to the exhaustive computation in PoW.
- **Practical Byzantine Fault Tolerance:** Practical Byzantine Fault Tolerance (PBFT) was the inaugural proposed consensus mechanism for addressing Byzantine fault tolerance, achievable even in a hostile node within the system. Executing non-deterministic chain codes can be quite effective. Stellar, Hyperledger, and Ripple exemplify PBFT.

- **Delegated Proof of Stake and Proof of Stake:** Delegated Proof of Stake (DPoS) and PoS differ fundamentally in that PoS operates as a representative democratic system, whereas DPoS functions as a direct democratic system. Users construct and validate blocks. The reduced number of nodes for block validation enhances the probability of rapid verification and accelerates transaction confirmation.
- **Proof of Authority:** Proof of Authority (PoA) integrates elements of PoW and PoS. It honours a stakeholder's reputation and identity. A shareholder is thereby bolstered by their reputation rather than their equity stake. Consequently, the foundational elements of the blockchain are safeguarded by authentic and reliable persons.
- **Proof of Vote algorithm:** The Proof of Vote (PoV) algorithm is unique compared with all other consensus mechanisms. To produce transaction blocks on the blockchain, a consortium of firms must participate in exchanging business data. As a result, they employ external personnel for assistance. The team will allocate the block to each firm inside the network to validate votes, ensuring the blockchain's decentralised nature. Business owners sometimes expand the workforce by employing individuals. This method was explicitly designed for a consortium.

1.2.5 Application of Blockchain Technology in Healthcare

Examples of blockchain applications in the healthcare sector include medical management, medical development and research genetics, and e-healthcare. Blockchain offers a secure and dependable mechanism for information storage across all healthcare industry sectors, enabling physicians and healthcare providers to use recorded data for various transactions and analyses [18,19]. In this section, we will examine alternative ways for implementing blockchain across different healthcare sub-domains, as seen in Figure 1.3. Blockchain technology transforms healthcare by addressing numerous challenges, including data security, cooperation, and efficiency. These are some of the key areas where blockchain is making a significant impact on the healthcare sector.

Medical Records Management: It is challenged by the fragmented nature of healthcare data, which often spreads across multiple providers. This disjointed setup limits access, complicates data sharing, and heightens security risks. Blockchain addresses these issues by creating a secure, decentralised ledger where patient records are cryptographically stored, making them accessible only through authorised channels. This setup enhances security, supports interoperability, and gives patients greater control over their health information. The past decade has seen significant growth in the digitalisation of health data, with healthcare professionals increasingly requiring immediate access to patient information to make timely decisions. Electronic health records (EHRs), digital versions of medical records, facilitate this need by allowing healthcare providers to access a patient's medical history, diagnoses, and treatment plans at any time, given proper authorisation. Blockchain's primary applications in EHRs include secure

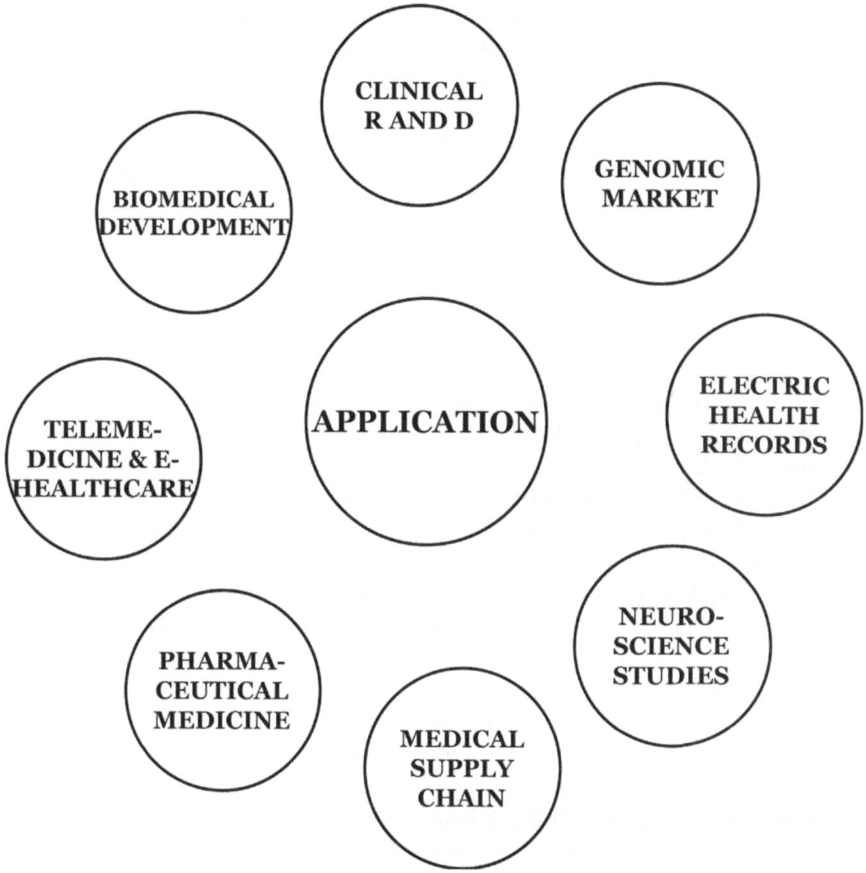

FIGURE 1.3 Blockchain technology implementation.

storage of prescription information, treatment protocols, laboratory results, test outcomes, and similar critical health data, thereby supporting a more integrated and secure healthcare data management system [20].

Supply Chain Management: This requires strict maintenance of the authenticity and integrity of medical equipment throughout the supply chain to ensure patient safety and compliance with regulatory standards. Blockchain enables transparent oversight across the supply chain, providing traceability from producer to consumer. Through smart contracts, verification procedures can be automated, ensuring that items meet quality standards at each stage and reducing the risk of counterfeit pharmaceuticals.

Clinical Trials: It relies on the integrity of trial data, transparency, and the management of participant consent to ensure research validity and adherence to ethical standards. Blockchain can streamline the administration of consent forms, securely archive trial data, and provide verifiable records of trial outcomes. This enhances transparency, reduces risks of data tampering, and expedites regulatory approvals.

Healthcare Data Sharing: It involves exchanging sensitive patient information among healthcare providers, researchers, and insurers, often complicated by privacy concerns and diverse IT systems. Blockchain allows patients to control access to medical information through encryption keys, enabling secure, patient-directed data sharing. By securely transmitting encrypted data, blockchain facilitates information flow for research and treatment while ensuring patient confidentiality.

Billing Information and Medical Claim Processing: Healthcare often involves multiple intermediaries, leading to inefficiencies, delays, and potential inaccuracies. Blockchain can streamline billing processes through smart contracts, reducing administrative costs, minimising fraud, and improving payment accuracy.

Identity Management and Authentication: Healthcare presents challenges in managing patient identities across different providers and systems, often resulting in inaccuracies and opportunities for fraudulent activities. Blockchain-based digital IDs provide a unified and secure approach to verifying patient identities, significantly enhancing treatment accuracy, billing, and patient engagement.

Research and Intellectual Property Protection: It is critical for ensuring innovation in healthcare. Safeguarding intellectual property and maintaining research integrity are essential components of this process. Blockchain enables the timestamping and authentication of research data and intellectual property rights, limiting unauthorised use. This technology fosters collaboration among researchers while protecting their rights and contributions.

Internet of Things (IoT) Integration: Healthcare involves securely transmitting real-time data to healthcare systems, which presents data integrity and confidentiality challenges. Blockchain provides a decentralised and secure platform for IoT devices to transmit this real-time data, ensuring its authenticity. In addition, it facilitates automated actions based on smart contracts, enhancing the efficiency and reliability of healthcare processes.

Various blockchain innovations in medical management highlight their applications, descriptions, and associated tools. In EHRs, blockchain creates a decentralised and immutable system that ensures patient data remain secure and tamper-proof using tools like MedRec, Patientory, and Guardtime Health. Blockchain enhances trial data transparency for clinical trials and research utilising platforms such as IBM Blockchain, FarmaTrust, and Embleema. In drug supply chain and management, blockchain technology tracks pharmaceuticals throughout the supply chain, effectively preventing counterfeit drugs, with tools like MediLedger, Blockpharma, and Chronicled. Regarding medical billing and claims, the blockchain validates the billing process, reducing fraud and errors; notable tools include PokitDok, Hashed Health, and SolveCare. Blockchain also plays a crucial role in patient consent management by securely recording and easily verifying patient consent for data sharing through tools like Consentium, Health Wizz, and Coral Health [21].

Furthermore, it facilitates interoperability by enabling data sharing across different healthcare systems while maintaining data integrity using technologies such as Hyperledger, Ethereum, and IBM Blockchain. For wearables and IoT devices, blockchain secures data from wearable health devices, ensuring privacy and accuracy with tools like Factom, Medicalchain, and Chronicled. In the insurance sector, blockchain enhances the transparency of claims processing, exemplified by Aetna's blockchain initiative and MetLife's LumenLab. In addition, blockchain provides highly secured digital identities for patients and medical experts through platforms like SelfKey, Civic, and ShoCard. Finally, blockchain improves public health data collection and reporting accuracy and speed in public health surveillance utilising tools such as Hacera, IBM Blockchain, and SAP Blockchain.

1.3 CHALLENGES IN BLOCKCHAIN ADOPTION

Blockchain technology faces challenges due to inherent defects and many architectural issues. As developed by various academics, blockchain technology has considerable technological problems and constraints, as illustrated in Figure 1.4.

The Problem of Throughput: Its maximum transaction rate does not exceed five. Consequently, if conventional e-commerce companies implement blockchain for transactions, it is likely to fail due to transaction delays and escalating network communication expenses.

Latency Issue: Elevated latency constitutes a significant issue inside the Bitcoin network. Procedures require approximately 10 minutes to complete, with additional time needed to avert duplicate spending. For the dual assault to succeed, certain criteria must be met.

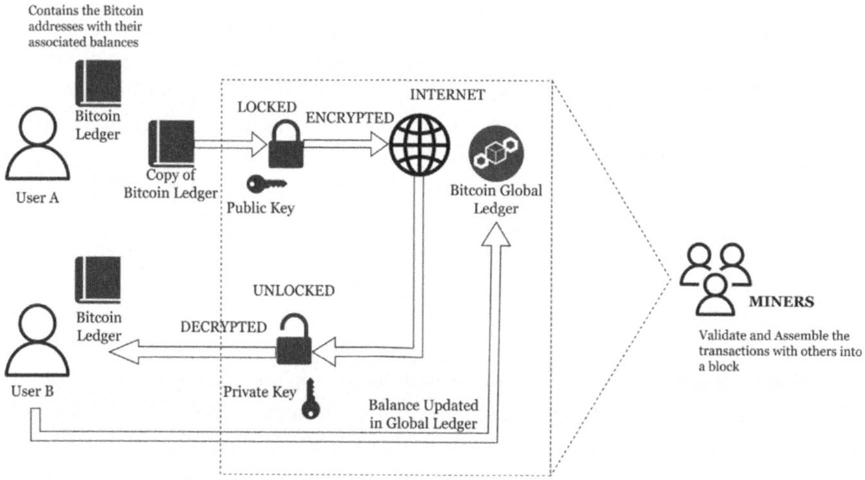

FIGURE 1.4 Challenges in blockchain technology.

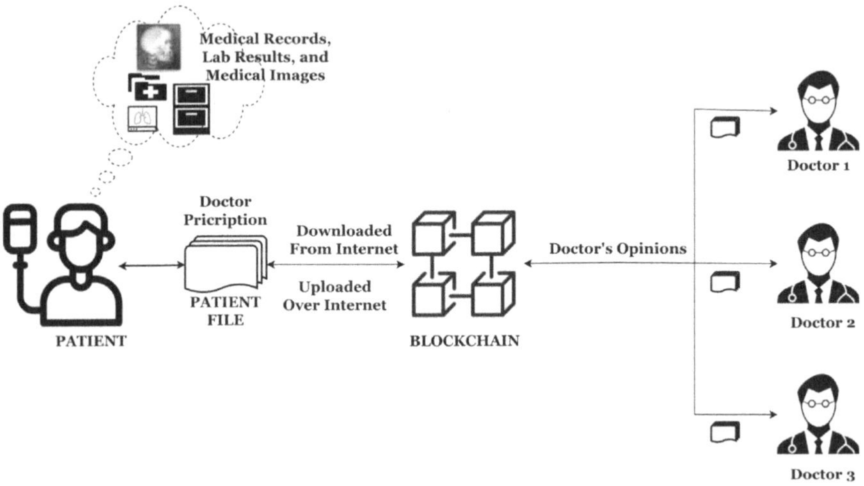

FIGURE 1.5 Scalability issue in blockchain technology.

Limitation of Size and Bandwidth: Blockchain scalability is essential for competing with conventional e-commerce platforms like Visa. The Bitcoin network can process approximately 500 transactions per second, facilitating scalability, as illustrated in Figure 1.5.

Block Size: Every node in a decentralised system processes each transaction, constituting a comprehensive replica of the ledger state. In the patient care system, only the essential ledgers for specific nodes are required, rather than the complete ledger.

High Volume: A million additional records may be replicated across all participants when this platform is employed to maintain patient profiles. Consequently, there would be an excess of information. Significantly, the creation of scaled-linked providers faces the difficulty of excessive data volume in healthcare facilities.

Transactions: Each transaction on the Ethereum blockchain requires verification by all validators inside the network. The data load adversely affects the network's speed. Consequently, achieving consensus or authenticating digital identification requires time.

Number of Nodes: The complexity of operating blockchain-based applications escalated with a substantial growth in users or patients. An increase in the overall computational requirements of the blockchain system accompanies this.

Protocol: Blockchain technology cannot address certain patient issues, such as unforeseen access or emergencies. Moreover, access to personal information by a physician was first prohibited during the emergency.

Resource Wastage: The consensus process is essential to the Bitcoin system, providing security against adversaries, but it also results in significant wastage of computational resources, manifesting as elevated hardware costs and energy consumption.

Usability Problem: Most blockchain research emphasises technical elements such as security, secrecy, data integrity, efficiency, and scalability, sometimes neglecting usability [22,23].

Fork Problem: A division occurs in the blockchain when a modification is required. Forks are categorised into two types: soft forks and hard forks. Blockchain operates through participants, nodes, and miners. Nodes must accept all modifications. In addition, updated nodes persist in validating blocks.

1.4 IMPLEMENTATION OF BLOCKCHAIN IN MEDICAL CARE

Implementing blockchain in medical care presents a transformative opportunity for healthcare institutions to enhance data security and transparency, ultimately improving patient care. The workflow commences with identifying specific use cases where blockchain can add value, such as medical records management, clinical trials, supply chain integrity, or insurance claims processing. After ascertaining these applications, the next step involves designing a blockchain network tailored to the organisation's needs. This includes selecting from private, public, or hybrid blockchain options based on data sensitivity and regulatory obligations and elucidating the consensus mechanisms and governance frameworks that will be employed. Once the network design is complete, healthcare data is gathered from diverse sources, including hospitals, clinics, and laboratories. Critical information is then encrypted using sophisticated cryptographic methods to ensure confidentiality.

Smart contracts are created to automate patient consent management, insurance verification, and medicine traceability. These smart contracts implement predetermined regulations and autonomously conduct transactions. As transactions occur, encrypted healthcare data and transaction records are preserved as immutable blocks on the blockchain. Each transaction, whether a patient visit or laboratory test result, is timestamped and linked to the preceding block, establishing a visible and verifiable information chain. Cryptographic key techniques maintain access control and privacy, ensuring that only authorised entities, including patients, healthcare providers, and researchers, can access designated data. This approach safeguards patient confidentiality and adheres to data protection laws such as General Data Protection Regulation (GDPR) and HIPAA.

Furthermore, interoperability is crucial for effective blockchain integration in healthcare; thus, standards and protocols must be developed to facilitate seamless data sharing among various healthcare systems and blockchain networks. Utilising Application Programming Interfaces (APIs) and data standards, such as HL7 and FHIR, enables smooth data interchange. Routine audits are essential to ensure data integrity and compliance with regulatory standards; the transparent and auditable nature of blockchain simplifies the auditing process and enhances regulatory compliance. Integrating blockchain with IoT devices and artificial intelligence algorithms further enhances the healthcare ecosystem by improving real-time data collection, analysis, and decision-making [24].

Continuous improvement is also vital; evaluating performance indicators like transaction velocity, scalability, and cost-efficiency allows for ongoing enhancements to the blockchain network and its protocols, thereby increasing efficiency and user experience [25]. Education and training for healthcare professionals, administrators, and patients regarding blockchain technology, its benefits, and the secure utilisation of blockchain-enabled applications are equally important [26,27]. Finally, fostering community engagement through collaboration among stakeholders, including healthcare providers, technology vendors, and regulators, can help address challenges and improve the implementation of blockchain technology in the healthcare sector.

1.5 CONCLUSIONS AND FUTURE SCOPE

This chapter delineates the current state of blockchain in healthcare administration, emphasising its prospective advantages and uses. Blockchain offers the potential to create safe, decentralised systems that emphasise patient-centred care through improved data processing and management capabilities. The suggested applications of blockchain in health data management encompass enhancing privacy, security, accessibility, and comprehensive control over EHRs. Various prototypes and concepts have been investigated, including payment systems, permissioned blockchains, and offline storage. Nevertheless, additional research is required to comprehensively comprehend and assess the efficacy of blockchain in healthcare. Furthermore, current projects must tackle challenges associated with their implementation, encompassing dependability, latency, and privacy difficulties. This sophisticated technique should be employed wisely, contingent upon its relevance and requirement for particular clinical situations.

REFERENCES

1. B. Haleem, M. Javaid, R. P. Singh, R. Suman, S. Rab, "Blockchain Technology Applications in Healthcare: An Overview," *International Journal of Intelligent Networks*, vol. 2, pp. 130–139, 2021. https://doi.org/10.1016/j.ijin.2021.09.005.
2. W. Y. Ng, T.-E. Tan, P. V. H. Movva, A. H. S. Fang, K.-K. Yeo, D. Ho, F. S. S. Foo, Z. Xiao, K. Sun, T. Y. Wong, A. T.-H. Sia, D. S. W. Ting, "Blockchain Applications in Health Care for COVID-19 and Beyond: A Systematic Review." *Lancet Digit Health*, vol. 3, no. 12, pp. e819–e829, 2021.
3. M. A. Bazel, F. Mohammed, M. Ahmed, "Blockchain Technology in Healthcare Big Data Management: Benefits, Applications and Challenges," In: 2021 *1st International Conference on Emerging Smart Technologies and Applications (eSmarTA)*, Sana'a, Yemen, August 10–12, 2021, pp. 1–8. https://doi.org/10.1109/eSmarTA52612.2021.9515747.
4. M. Y. Jabarulla, H. N. Lee, "Blockchain-Based Distributed Patient-Centric Image Management System," *Applied Sciences*, vol. 11, no. 1, p. 196, 2021. https://doi.org/10.3390/app11010196.
5. A. Tandon, A. Dhir, A. K. M. Najmul Islam, M. Mäntymäki, "Blockchain in Healthcare: A Systematic Literature Review, Synthesising Framework and Future Research Agenda," *Computers in Industry*, vol. 122, p. 103290, 2020. https://doi.org/10.1016/j.compind.2020.103290.
6. H. D. Zubaydi, Y.-W. Chong, K. Ko, S. M. Hanshi, S. Karuppayah, "A Review on the Role of Blockchain Technology in the Healthcare Domain," *Electronics*, vol. 8, p. 679, 2019. https://doi.org/10.3390/electronics8060679.

7. G. Chen, B. Xu, M. Lu, N. S. Chen, "Exploring Blockchain Technology and Its Potential Applications for Education," *Smart Learning Environments*, vol. 5, p. 1, 2018. https://doi.org/10.1186/s40561-017-0050-x.
8. H. F. Anjum, S. Z. A. Rasid, H. Khalid, M. Alam, S. M. Daud, H. Abas, "Mapping Research Trends of Blockchain Technology in Healthcare," *IEEE Access*, vol. 8, pp. 174244–174254, 2020. https://doi.org/10.1109/ACCESS.2020.3025011.
9. N. Singh, Y. Kumar, "Security and Privacy Requirements in 5G Healthcare," In: *Block Chain for 5G Healthcare Applications*, IET Publication, pp.1–30, 2021. https://doi.org/10.1049/PBHE035E_ch1.
10. Al Mamun, A. S. Azam, and C. Gritti, C. "Blockchain-Based Electronic Health Records Management: A Comprehensive Review and Future Research Direction," *IEEE Access*, vol. 10, pp. 5768-5789, 2022. https://doi.org/10.1109/ACCESS.2022.3141079.
11. H. Kaur, M. A. Alam, R. Jameel, A. K. Mourya, V. Chang, "A Proposed Solution and Future Direction for Blockchain-Based Heterogeneous Medicare Data in Cloud Environment," *Journal of Medical Systems*, vol. 42, no. 8, p. 156, 2018. https://doi.org/10.1007/s10916-018-1007-5.
12. D. Elangovan, C. S. Long, F. S. Bakrin, C. S. Tan, K W. Goh, S. F. Yeoh, M. J. Loy, Z. Hussain, "The Use of Blockchain Technology in the Health Care Sector: Systematic Review," *JMIR Medical Informatics*, vol. 10, p. e17278, 2022. https://doi.org/10.2196/17278.
13. N. Goenkaa, A. K. Sharmab, S. Tiwaric, N. Singhd, V. Yadave, S. P. K. Chadaga, "A Regularized Volumetric ConvNet Based Alzheimer Detection Using T1-Weighted MRI Images," *Cogent Engineering*, vol. 11, no. 1, p. 2314872, 2024. https://doi.org/10.1080/23311916.2024.2314872.
14. I.A. Omar, M. Debe, R. Jayaraman, K. Salah, M., Omar, J. and Arshad, "Blockchain-Based Supply Chain Traceability for COVID-19 Personal Protective Equipment. *Computers and Industrial Engineering*, vol. *167*, p.107995, 2022. https://doi.org/10.1016/j.cie.2022.107995.
15. L. Kandasamy, A. Ajay, "Implementation of Blockchain Technology for Secure Image Sharing Using Double Layer Steganography," In: Hu, Z., Wang, Y., He, M. (eds), *Advances in Intelligent Systems, Computer Science and Digital Economics IV.* CSDEIS 2022. Lecture Notes on Data Engineering and Communications Technologies, vol 158. Springer, pp. 186–195, 2023. https://doi.org/10.1007/978-3-031-24475-9_16.
16. R. Pandey, A. Khatri, G. Premalatha, M. Lakshmu Naidu, D. Verma, C. J. Shinga Diya, "Integration of Blockchain Technology in Health Care System: An Intensive Review," In: *2023 3rd International Conference on Advance Computing and Innovative Technologies in Engineering (ICACITE)*, Greater Noida, India, May 12–13, 2023, pp. 968–972. https://doi.org/10.1109/ICACITE57410.2023.10182960.
17. E. Elgamal, W. Medhat, M. A. Elfatah, N. Abdelbaki, "Blockchain in Healthcare for Achieving Patients' Privacy," In: *2023 20th Learning and Technology Conference (L&T)*, Jeddah, Saudi Arabia, Jeddah, Saudi Arabia, January 26, 2023, pp. 59–64. https://doi.org/10.1109/LT58159.2023.10092352.
18. V. Asha, A. P. Nirmala, B. Saju, A. Sk, A. K. Das, A. Sharma, "Role of Blockchain Technology in Healthcare," In: *2023 International Conference on Sustainable Computing and Data Communication Systems (ICSCDS)*, Erode, India, March 23–25, 2023, pp. 370–374. https://doi.org/10.1109/ICSCDS56580.2023.10104832.
19. D. C. Nguyen, M. Ding, P. N. Pathirana, A. Seneviratne, "Blockchain and AI-Based Solutions to Combat Coronavirus (Covid-19)-Like Epidemics: A Survey," *IEEE Access*, vol. 9, pp. 95730–95753, 2021. https://doi.org/10.1109/ACCESS.2021.3093633.
20. M. K. Chandol, M. Kameswara Rao, "Blockchain-Based Cryptographic Approach for Privacy Enabled Data Integrity Model for IoT Healthcare," *Journal of Experimental & Theoretical Artificial Intelligence*, vol. 37, 53–74. https://doi.org/10.1080/0952813X.2023.2183268.

21. R. K. Dewang, M. P. Yadav, S. Awasthi, O. Raj, A. Mewada, K. L. Bawankule, "Data Secure Application: An Application That Allows Developers to Store User Data Securely Using Blockchain and IPFS," *Multimedia Tools and Applications*, vol. 83, no. 15, pp. 45491–45517, 2024.
22. S. Upadhyay, S. Agarwal, A. Chaudhary, P. Singh, R. Sharma, "Blockchain in Healthcare Systems: A Systematic Review," In: *2024 2nd International Conference on Disruptive Technologies (ICDT)*, Greater Noida, India, March 15–16, 2024, pp. 140–147. https://doi.org/10.1109/ICDT61202.2024.10489533.
23. A. Arif, M. Hussain, C. P Subbe, "Blockchain: What Is the Use Case for Physicians in 2024? A Rapid Review of the Literature," *Future Healthcare Journal*, vol. 11, no. 1, p. 100005, 2024. https://doi.org/10.1016/j.fhj.2024.100005.
24. M. S. B. Kasyapa, C. Vanmathi, "Blockchain Integration in Healthcare: A Comprehensive Investigation of Use Cases, Performance Issues, and Mitigation Strategies," *Frontiers in Digital Health*, vol. 6, p. 1359858, 2024.
25. M. A. Ansari, D. K. Singh, "Human Detection Techniques for Real Time Surveillance: A Comprehensive Survey," *Multimedia Tools and Applications*, vol. 80, pp. 8759–8808, 2021. https://doi.org/10.1007/s11042-020-10103-4.
26. A. Bathula, S. K. Gupta, S. Merugu, L. Saba, N. N. Khanna, J. R. Laird, S. S. Sanagala, R. Singh, D. Garg, M. M. Fouda, J. S. Suri, "Blockchain, Artificial Intelligence, and Healthcare: The Tripod of Future—A Narrative Review," *Artificial Intelligence Review*, vol. 57, p. 238, 2024.
27. K. H. Kim, S.-M. Kim, Y. M. Park, E. S. Lee, S. J. Jung, J. Kang, D. U. An, K. Min, S. R. Shim, H. W. Yu, H. W. Han, "A Blockchain-Based Healthcare Data Marketplace: Prototype and Demonstration," *JAMIA Open*, vol. 7, no. 2, p. ooae029, 2024.

2 Towards a Unified Framework for Blockchain-IoT Integration

Architectural Insights and Research Challenges

Reetu Singh, Greetta Pinheiro, Niharika Keshari, Arvind Mewada, and Mohd. Aquib Ansari

2.1 INTRODUCTION

The Internet of Things (IoT) represents a new configuration that alters the way people interact with their environment. It is an incredibly complex network of internet-connected devices, ranging from simple appliances used in the home to highly complex industrial equipment. IoT devices collect data, exchange information, and utilise software, sensors, and connectivity to automate processes and enable real-time decisions to be made to improve efficiency and convenience in work and daily life [1]. IoT is creating intelligent environments. The smart thermostat can learn your preferences, and heavy industrial equipment is capable of sensing when it is about to break down. Every device in the network generates a huge amount of data, so we need systems that can store them efficiently for processing and analysing without sacrificing performance. In addition, as the data increases, so do concerns about security, immutability, privacy, and the need for reliable, scalable infrastructure. Blockchain technology with core attributes—security, immutability, and transparency—emerged as a feasible solution to the significant challenges of IoT [2]. First designed as the backbone of cryptocurrencies like Bitcoin, blockchain is a decentralised [3] ledger system that securely records and stores data over a distributed network. Blockchain works by recording the information and the hash of the previous information into encrypted "blocks," each linked together to form an unalterable and secure chain. Any change to the blockchain requires consensus from the entire network, thus affording strong protection against forgery or unauthorised changes. This decentralised architecture lowers the risk of reliance on some central authority for transfers of data, thereby minimising the risk of a single point of failure, which is particularly beneficial for IoT networks where many devices exchange sensitive information [4].

DOI: 10.1201/9781003545620-3

Integrating blockchain and IoT provides a revolutionary solution to some of the most significant hurdles against security and scalability. Each transaction in blockchain-based IoT networks is encrypted and validated by a distributed network, thus making the system highly secure against hacking or data corruption [4]. For this reason, blockchain-based IoT allows the independent and secure communication of devices because data can be shared with minimal centralised control [5]. This makes IoT networks more efficient and also resilient to tampering and unauthorised access [4]. Apart from security, blockchain addresses the scalability challenges IoT faces as it expands. The levels of data generated by IoT present a challenge to efficiently managing traditional centralised systems. This results in inefficiency and delay in such systems.

With a decentralised architecture, blockchain ensures that the workload is distributed across many nodes, thus making the system scalable. As IoT networks expand, blockchain offers them high performance and reliability without the data processing load that might be congested with it. Smart contracts present much more than increased security and scalability: they tremendously boost the potential offered by IoT in blockchain frameworks. As conditions are integrated directly into the blockchain, smart contracts automatically execute conditions while allowing IoT devices to automate transactions and processes without any human intervention. The smart contracts in a smart city, for example, control how the autonomous vehicles, traffic sensors, and traffic lights interact. They optimise the flow of traffic and reduce congestion in cities. They also automate the payment disbursement or inventory handling on some predefined criteria.

This streamlines operations. Hence, blockchain in supply chain management creates a transparent and secure system for tracing products from their source to the end customer, thereby reducing fraud and enhancing traceability and accountability. Monitoring of patient health can be done by IoT devices while ensuring that blocks within this system secure the integrity and privacy of the patient's data, and access will only be granted to parties authorised to view it [6]. The blockchain-based innovation of IoT shapes future interaction between devices and systems, propelling sectors towards innovation, automation, and efficiency using trustworthy solutions for industries and consumers alike.

This chapter dives into IoT and blockchain architectures, their benefits when combined, as well as challenges involved. A rigorous literature review of existing research work will establish key points of gaps and suggest future directions to improve blockchain-based IoT. The chapter is organised as follows: In Section 2.2, the basics of IoT and blockchain are dealt with, and then the entire architecture of IoT-blockchain with its applications and challenges are addressed in Section 2.3. In Sections 2.4–2.6, related research is reviewed, gaps are pointed out, future research directions are proposed, respectively, and a conclusion is included in Section 2.7.

2.2 PRELIMINARIES

2.2.1 Internet of Things (IoT)

It is described as the interconnection of numerous networks of devices with sensors, software, and internet tools, enabling those appliances to gather, share, and process information. The origin of IoT is in extending the internet beyond its traditional computing devices to a wide range of objects that make more interactions and automation in our daily lives. The IoT system consists of several components, such as sensors and actuators, communication protocols, data processing systems, and applications that allow users to access and remotely control these devices [7]. This network allows improvements in efficiency and innovation in many sectors. There are other industries, such as smart homes, health, manufacturing, and transport, which shall benefit from real-time monitoring and the control of processes by the IoT. Typically, three primary layers constitute the architecture of IoT:

1. **Perception Layer:** Within that layer, IoT devices, such as smart appliances, cameras, and a wide variety of sensors, capture data about the physical environment. Its primary purpose is to sense, monitor, and obtain information from the environment.
2. **Network Layer:** The network layer uses several protocols, such as Wi-Fi, Bluetooth, and cellular networks, to offer communication between devices and data processing systems that assure the secure transmission of information to allow collected information to be relayed efficiently to the proper applications or storage systems.
3. **Application Layer:** This layer includes the software applications that analyse collected data so that meaningful insights or tools to control can be presented to the users. It supports automation and enables better decision-making.

2.2.2 Blockchain

Blockchain technology is a decentralised and distributed digital ledger, allowing the secure capture of transactions across several computers and providing assurance of data security, immutability, and transparency. While it was first designed for cryptocurrencies like Bitcoin, the primary idea of blockchain lies in its capacity to generate a secure, tamper-resistant record of transactions without needing a central authority [8,9]. Transactions are grouped into blocks subsequently connected to previous blocks, resulting in a secure chain, as illustrated in Figure 2.1. The transactions are hashed and combined in pairs to form a tree-like structure where each

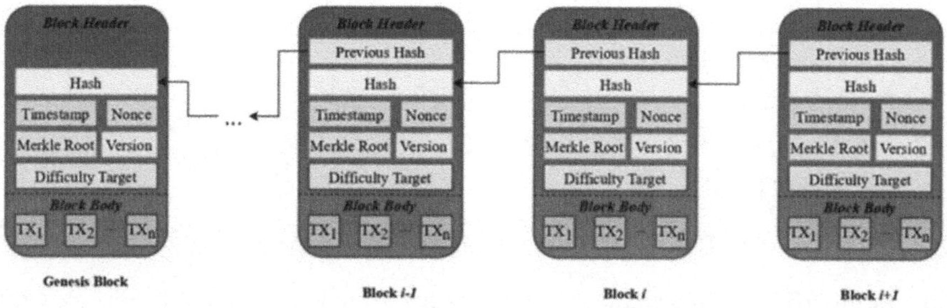

FIGURE 2.1 Blockchain architecture: Each block contains a header with a previous block hash, nonce, timestamp, and Merkle root, forming a chain of blocks.

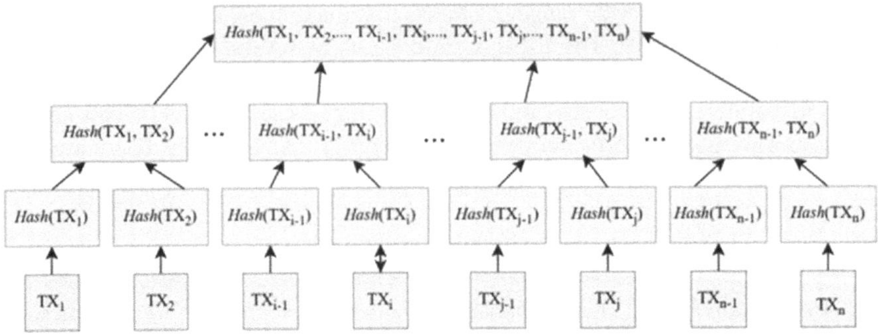

FIGURE 2.2 Merkle tree and hashing in blockchain.

node represents the hash of its child nodes, as shown in Figure 2.2. The block header includes the root hash, ensuring transaction integrity and security. This architecture promotes trust among participants, as any alteration in the data would require consensus from the entire network.

Blockchain architecture is defined by its decentralised and distributed characteristics, consisting of several key components:

1. **Blocks:** Every block holds the transaction data, timestamp, and cryptographic hash of the previous block, effectively linking them into a secure chain (as shown in Figure 2.1). In addition, blocks utilise Merkle Trees for efficient and secure verification of transactions, ensuring data integrity (depicted in Figure 2.2). This structure ensures the integrity and immutability of the recorded data.
2. **Nodes:** In a blockchain network, individual computers are called nodes; each holding a copy of the blockchain is involved in validating transactions. This decentralised model prevents any single entity from controlling the entire system, thereby improving the security and transparency of the system (illustrated in Figure 2.3). Each node verifies transactions and maintains the state of the Blockchain, contributing to the resilience and reliability of the network.
3. **Consensus Mechanisms:** Protocols are used to ensure agreement on transaction validity. Popular methods include Proof of Stake (POS) and Proof of Work (PoW), which help maintain data integrity, eliminating the need for a central authority. The consensus process and the formation of new blocks are depicted in Figure 2.3.

2.3 IOT USING BLOCKCHAIN: ARCHITECTURE, ADVANTAGES, APPLICATIONS, CHALLENGES

2.3.1 IoT Using Blockchain: Architecture

Integrating IoT and blockchain requires a multi-layered architecture to ensure seamless operation, communication, security, and data integrity across devices, networks, and services. The following section describes the layered architecture of IoT systems

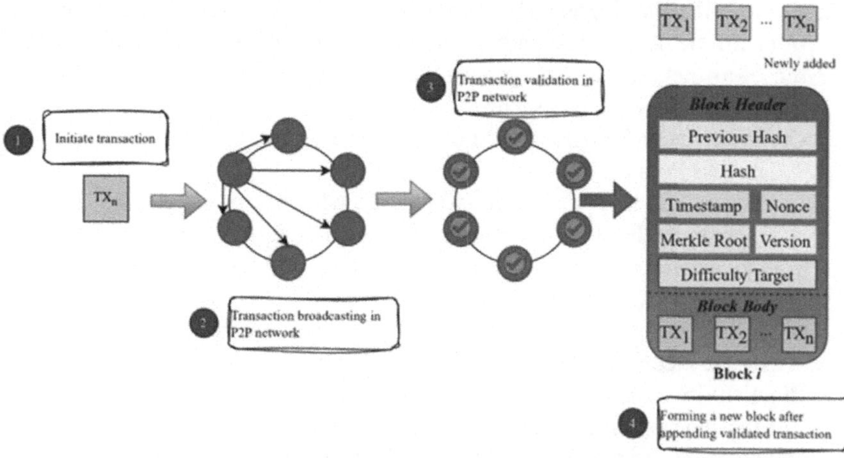

FIGURE 2.3 Blockchain transaction flow: (1) Initiating a transaction, (2) Transaction broadcasting in the P2P network, (3) Transaction validation, and (4) Formation of a new block.

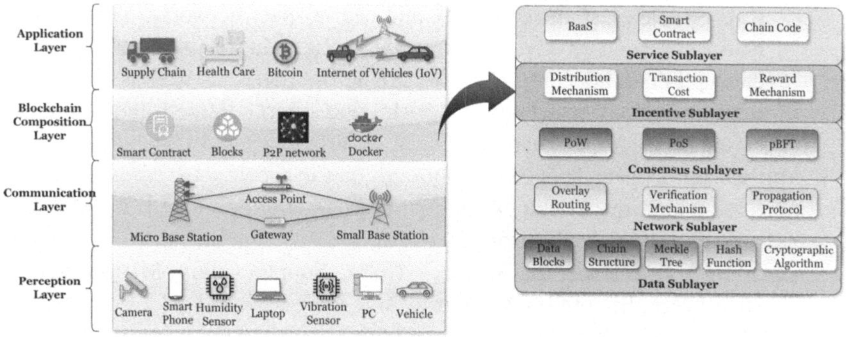

FIGURE 2.4 Layered architecture of IoT using blockchain.

using blockchain based on Figures 2.3 and 2.4 illustrates how blocks are linked and how transactions are validated and added to the blockchain.

1. **Perception Layer:** The perception layer, also known as the device layer, consists of the physical devices that collect data from the environment. These devices are embedded with sensors, actuators, and IoT technologies that capture real-time information.
   Components: Cameras, Smartphones, Humidity Sensors, Vibration Sensors, Laptops, PCs, Vehicles.
   Function: This layer acts as the interface that translates the real world into the digital system. This layer gleans environmental data, such as humidity, temperature, or video feeds and motion, and sends it to the communication layer.

2. **Communication Layer:** The communication layer is used for data transfer from IoT devices to the perception layer of the blockchain network. The sensor-to-the-gateway and gateway-to-the-processing unit communication is absolutely reliable and secure.
   Components: Micro Base Stations, Gateways, Small Base Stations, and Access Points.
   Function: The layer discusses the data transference protocols, which further states that in order to transfer information from the devices to the blockchain system, communication has to be wireless. It is a case of having network components involved, namely, base stations, gateways, and access points, which will facilitate electronic data communication by the electronic devices to the central servers or cloud platforms.
3. **Blockchain Composition Layer:** Block chain layer for composition concerning blockchain is important for safe, immutable, and distributed data management. It includes processing and storing IoT data by using a decentralised ledger.
   Components: Blocks, Smart Contracts, Docker Containers, P2P (Peer-to-Peer) Networks.
   Function: This layer marks the area of application of blockchain's decentralised ledger, smart contracts, and peer-to-peer networks. The blockchain composition layer ensures the protection of IoT data through encryption and decentralised storage across nodes, as well as automating transactions or data exchange through smart contracts and containerised applications, such as Docker, for efficiency and scalability.
4. **Application Layer:** This layer elaborates on the specific use cases and services for which the IoT-blockchain system is being developed. It links an end user to various applications spread across different industries.
   Components: Internet of Vehicles (IoV), Supply Chain Management, Healthcare, Bitcoin Integration.
   Function: Blockchain-based IoT solutions specifically meet the needs of varied industries, such as traceability in supply chains, health monitoring, vehicle-to-vehicle communication, and cryptocurrency transactions.
5. **Service Sublayer (from the right-hand portion of the diagram):** This service layer provides Blockchain-as-a-Service, which allows for the execution of smart contracts and chain code. It provides the basic services that are required to design and deploy blockchain applications.
   Components: Smart Contracts, BaaS, Chain Code.
   Function: It ensures the seamless working of the decentralised application with the BaaS frameworks given to the IoT developers who help develop blockchain technologies.
6. **Incentive Sublayer:** The incentive sublayer provides an incentive sublayer, ensuring economical operations inside the IoT-blockchain system.
   Components: Distribution Mechanism Transaction Cost Reward Mechanism.
   Function: It deals with the incentive mechanisms by distributing rewards to the participants or nodes inside the blockchain network so that they can use their computing power to validate transactions.

7. **Consensus Sublayer:** This sublayer is crucial for upholding the integrity of the blockchain network through consensus algorithms.
   Components: Practical Byzantine Fault Tolerance, PoS, PoW.
   Function: A consensus sublayer verifies that all parties agree about the legitimacy of transactions added to the blockchain. There are several consensus algorithms, including PoW, PoS, and pBFT, which serve to secure the validation of transactions and prevent fraudulent acts.
8. **Network Sublayer:** A network sublayer supports secure routing and propagation of data across the Blockchain.
   Components: Verification Mechanisms, Overlay Routing, Propagation Protocols.
   Function: This layer emphasises secure forwarding and validation of data packets throughout the network. Overlay routing techniques allow effective movements of blockchain data through the IoT network. The verification of true data is guaranteed as the overlay ensures validation and back-checks of data while communicating between sources and destinations.
9. **Data Sublayer:** This sublayer of the blockchain forms the data structure and encryption in the blockchain.
   Components: Blocks, Chain, Merkle Tree, Hash Functions, Cryptographic Algorithms,
   Function: This sublayer organises data into blocks, connects them, and locks the blocks using cryptographic methods by Merkle trees and hash functions. The result is that the data coming through the IoT is tamper-proof, verifiable, and will be secured on the distributed ledger.

2.3.2 Advantages of IoT Using Blockchain

The combination of IoT and blockchain combines the strengths of both to solve crucial challenges that exist in IoT systems, from security and scalability to trust. Here are the main benefits that come from combining IoT with blockchain.

2.3.2.1 Enhanced Security

Blockchain provides a decentralised and tamper-proof framework for the securement of IoT data [10]. The critical dangers associated with IoT, as these devices generate and share huge amounts of sensitive information, are susceptibility to cyberattacks. Blockchain has mitigated the risks through data integrity, which guarantees that whenever the data from the IoT devices are written to the ledger, it cannot be altered. All the attempts at changing the data shall be rejected by the network, leaving that authenticity intact. Decentralised validation rather than only relying on a central authority to verify and record transactions, blockchain utilises a series of nodes within a network. This alone is more resistant to attacks such as a Distributed Denial of Service attack on IoT networks.

2.3.2.2 Trustless Environment

In the traditional setups of IoT, data validation is carried out most of the time by a trusted central authority or third party. This, however, opens up vulnerabilities and complicates configurations. The application of blockchain in IoT systems allows the possibility of establishing a trustless environment where, theoretically, devices can self-governally interact and share data without intermediaries.

1. **Decentralised Trust:** The blockchain consensus mechanism allows IoT devices to validate and confirm transactions and data independently. Therefore, a trusted intermediary is no longer required, and this enhances efficiency and security.
2. **Machine-to-Machine (M2M) Trust:** Blockchain provides for secure and trustless communication between IoT devices so that M2M transactions can be effected safely, without any human intervention.

2.3.2.3 Improved Transparency and Traceability

The most important benefit of blockchain is its transparent and verifiable record of all interactions and transactions in an IoT network, which is especially useful in supply chain management, healthcare, and agriculture.

1. **Traceability:** The blockchain supply chain management system allows the stakeholders to trace the origin and journey and the current status of every product at each point. The entire transparency blockchain ensures that all stakeholders have the same information and fraud cannot occur.
2. **Auditability:** All the actions from an IoT device are straightforwardly registered on blockchain and can be audited. This allows users to track everything and adhere to the regulations or standards in real-time.

2.3.2.4 Automation Through Smart Contracts

Smart contracts are self-executing agreements that activate when predefined conditions written into their code are satisfied. Smart contracts further simplify IoT interactions amongst machines, ensuring transactions are possible without any human interface.

1. **Autonomous Operations:** Through real-time information, IoT devices can automatically initiate actions. For instance, a smart contract may be activated to automatically turn on irrigation in agriculture if the moisture level falls below specified parameters in the soil.
2. **Efficient Processes:** Smarter contracts are those that make payments, update records of inventory, or send orders automatically to be executed when pre-set conditions such as delivery confirmation are satisfied, thus reducing the time spent doing these things manually and hence making supply chains more efficient.
3. **Cost Saving:** Smart contracts automate the transaction and leave no need for an intermediary or oversight in an endeavour. Thus, it saves costs and allows transactions to be performed quickly.

2.3.2.5 Scalability and Resource Optimisation

Spreading out the workload to multiple nodes could be another way in which the decentralised architecture of blockchain would help improve the scalability of IoT systems because it would not bottle up with data.

1. **Efficient Resource Usage:** IoT systems create massive data sets, which must be processed and validated. Blockchain allows the network to distribute the task among various nodes so that no single point of the network is burdened by the same.
2. **Scalable Networks:** With an increase in the number of IoT devices, blockchain will be able to handle more transactions without such bottlenecks as in centralised systems that collapse when given too much information.

2.3.3 Applications of IoT Using Blockchain

Combining IoT and blockchain establishes a robust technological framework to revolutionise multiple industries. Blockchain helps address key challenges in IoT, such as transparency, security, scalability, and data integrity, making it ideal for numerous applications. Below are some significant applications of IoT using blockchain across different sectors:

2.3.3.1 Supply Chain Management

In supply chain management, IoT devices like sensors, Radio-Frequency Identification (RFID) tags, and GPS technologies are used to track goods and monitor conditions (temperature, humidity, etc.) throughout the complete life cycle of a product [11]. Incorporating blockchain into this process improves traceability and transparency.

1. **Real-Time Tracking:** IoT devices monitor the location and condition of goods in real-time, while blockchain provides a secure, tamper-proof record of these transactions.
2. **Tamper-Proof Records:** The use of blockchain ensures that each stage of the supply chain is documented on a secure, decentralised ledger. This allows stakeholders to verify and evaluate the condition of goods at every point, from their origin to their destination.
3. **Fraud Reduction:** The use of transparency and immutability by blockchain will reduce the risk associated with fraud, enhancing authenticity and trust in the supply chain.

2.3.3.2 Smart Cities

This makes it use IoT technology to control and maximise the efficiency of cities' systems, from the level of energy usage and its flow up to the availability of water and public safety [12]. This makes the idea of using blockchain to improve the security and efficiency of following IoT applications.

a. **Traffic Management:** The IoT sensors can track traffic conditions in real-time. The smart contracts on blockchain can alter the working of the traffic lights or usage of roads based on live data fed into it. Blockchain guarantees the genuineness of this data and automates actions.
b. **Energy Management:** Blockchain enables decentralised peer-to-peer energy trading, in which smart contracts allow residents to either sell or acquire surplus energy coming from photovoltaic solar panels or other

renewable energy sources. The byproduct is the powerful establishment of an efficient energy market within smart cities and limiting dependence on centralised utilities.

c. **Public Safety:** The IoT-based surveillance and sensors track environmental conditions. Blockchain may record and manage these data to make sure public safety decisions are not based on erroneous or false information.

2.3.3.3 Healthcare

IoT devices, such as smart medical equipment and surveillance systems at a distance, collect and transmit patient data in the health sector. Blockchain can then ensure that such sensitive information is secure, private, and accurate [13].

a. **Medical Data Security:** Blockchain technology ensures the secure transferability of medical records among service providers, patients, and insurers. With IoT devices generating real-time health information on an ongoing basis, blockchain stores it safely and cannot be modified.
b. **Remote Patient Monitoring:** Wearable IoT devices can enable remote patient monitoring by reporting patient vital signs such as heart rate, blood pressure, and glucose levels. Blockchain ensures safe data transmission; the data will only be accessible to authorised people, securing the privacy of the patient.
c. **Drug Traceability:** Blockchain and IoT may interlink to track the medicines at each stage of the drug supply chain from manufacturing to the point of sale, thereby reducing counterfeit drugs and ensuring safe, authentic drugs to the patients.

2.3.4 Agriculture and Food Supply Chain

Blockchain and IoT are of considerable importance to agriculture, aiming to improve food safety while optimising the use of resources, thus reducing waste.

1. **Precision Agriculture:** It is used to monitor soil, weather, and crop health conditions all through IoT sensors that store data safely in the blockchain. This offers better bases for farmer decisions on irrigation, planting, and harvesting that tend to increase yields and save resources.
2. **Food Safety and Traceability:** IoT will track the product from farm to consumer. Blockchain will ensure the right information, to the consumer who is consuming it in order to verify the origin and quality of the food consumed.
3. **Sustainability and Fair Trade:** Blockchain tracking at each step of production and distribution in an agricultural process helps to verify the practice of sustainability and fair trade in agriculture and that ethical standards are respected.

2.3.5 Automotive Industry and Autonomous Vehicles

Blockchain and IoT will have important effects on improving the functionality and security of connected and autonomous vehicles for higher efficiency.

a. **Vehicle-to-Everything (V2X) Communication:** The IoT sensors enable vehicles to get in touch with other vehicles, infrastructures, and pedestrians. Blockchain ensures that these communications remain secure and unaltered. This would be a critical feature for vehicle safety and reliability in autonomous driving.
b. **Ownership and Usage Tracking:** Blockchain can keep a record of ownership, maintenance, and usage of a vehicle in a safe manner. It is very useful for car-sharing and ride-hailing applications because user transparency is very critical in such scenarios.
c. **Insurance and Smart Contracts:** Data on an accident can be automatically collected through IoT sensors in a vehicle, and smart contracts on the blockchain can be drafted to automate processing for an insurance claim.

2.4 LITERATURE SURVEY

This literature survey provides an overview of key contributions, limitations, and future directions regarding integrating blockchain technology with the IoT, focusing on healthcare applications. The survey facilitates a clear comparison across the existing literature by synthesising findings from various studies. The integration of blockchain and IoT has emerged as a significant area of research, addressing various applications and challenges across multiple sectors. Article [8] examines this intersection within the healthcare sector, exploring applications such as remote patient monitoring, disease prediction, and secure medical record management. They highlight various blockchain types and associated tools that could enhance healthcare delivery. However, their review is confined to a specific timeframe (2018–2023) and focuses predominantly on healthcare, leading to a lack of practical implementations and reliance on conceptual rather than empirical solutions. They recommend developing more efficient blockchain algorithms to manage real-time healthcare data and enhance privacy. A comprehensive IoT and blockchain integration analysis using Latent Dirichlet Allocation to identify research themes across 4,455 journal articles [14]. They group those articles into 14 different research areas, but they are mainly theoretical, with very little emphasis on practical aspects. They have identified quite a number of relatively unexplored topics that may be included in future research, such as energy efficiency on wireless sensor networks and access control mechanisms in IoT systems. Article [15] analysed the IoT-blockchain architecture and challenges, focusing on several advantages, such as data integrity and security, with some potential future research challenges. Their work is mainly theoretical and not experimentally proven; the context is quite challenging in the development of robust frameworks to move forward with Blockchain Internet of Things (BIoT) security and overcome the challenge of scalability. In addition, [7] discuss various applications, opportunities, and challenges of the integration of IoT and blockchain over the diverse domains of healthcare, agriculture, and smart cities. Even though their framework is insightful, it is not yet equipped to address scalability concerns sufficiently, and only little empirical material supports their claims. They further remark that research in this area is necessary to boost the interoperability as well as security aspects in complex IoT-blockchain environments. Authors [16] take into consideration the way blockchain could ease IoT security issues considering the interaction of artificial intelligence (AI) with the two paradigms. They introduce the concept of blockchain as a service for IoT, but their work mainly presents theoretical models without empirical results or evaluations. Their future directions include investigating AI-based optimisation strategies for IoT and blockchain integration and further real-world deployments of blockchain as a service. Article [17] systematically reviews blockchain solutions for IoT, focusing on technical approaches for blockchain integration in IoT data and device management. However, much of the research remains at the conceptual stage, lacking real-world applications and facing high computational costs in large-scale IoT environments. They recommend developing scalable, lightweight blockchain frameworks to support large IoT systems better. The literature reveals a dynamic and evolving landscape for integrating blockchain and IoT. While significant theoretical contributions have been made, most studies emphasise conceptual frameworks with limited empirical validation. Future research should prioritise practical implementations, scalability solutions, and exploration of under-researched areas to harness the potential of blockchain and IoT integration fully.

2.5 RESEARCH GAPS IN IOT USING BLOCKCHAIN

2.5.1 Scalability Challenges

The low transaction throughput and latency in blockchain make it hard to handle the large amount of data generated by IoT devices in real-time. Scalability is very significant in IoT systems because this involves large numbers of devices and, thereby, results in an enormous quantity of data. With these scalability issues dealt with, using blockchain in large-scale IoT systems is feasible.

2.5.2 Storage and Data Management

In order to store large sets of data generated by IoT devices, blockchains must be developed efficiently. The requirement to carry all IoT data on-chain becomes unmanageable due to enormous storage requirements, which further causes problems such as blockchain bloat and cost augmentation. Thus, the development of efficient solutions for storing IoT data and handling it securely is of extreme importance without a significant load on blockchain systems.

2.5.3 Energy Consumption and Resource Constraints

However, traditional consensus mechanisms, such as PoW, require a lot of processing power and resources to work for IoT devices. Thus, the traditional blockchain consensus mechanisms are impractical for use on a massive scale in IoT networks. PoW is the most computer-intensive. Most IoT devices cannot provide such resource exhaustiveness due to their dependence on minimal power sources, like batteries. Hence, energy-friendly solutions become critical for the adoption of blockchain in IoT. It can extend blockchain adoption to even resource-constrained devices so long as the IoT devices have lightweight consensus mechanisms tailored for them, such as PoS or other low-power algorithms, improving security without losing efficiency.

2.5.4 Latency and Real-Time Processing

The delay in blockchain technology is due to the time that is being taken for block validation and consensus. Hence, it cannot be applied directly to real-time IoT applications such as autonomous vehicles and smart healthcare. Slight delays in such an environment can lead to major issues such as accidents or delayed medical interventions; thus, latency should be minimised for the implementation of blockchain in such applications. Low-latency blockchain solutions would enable IoT systems to benefit from the security and transparency of blockchain while still fulfilling real-time requirements of critical operations.

2.5.5 Security and Privacy Concerns

IoT devices remain vulnerable to cyberattacks, whereas blockchain enhances the security aspect. It is certainly not the end; it leaves open Distributed Denial of Service attacks or vulnerabilities in smart contracts. IoT systems are bound to hold sensitive information often, so the right security and privacy mechanisms must be there. No matter what Blockchain promises, it cannot solely assure the absoluteness of safety for the ecosystems of IoT against emerging threats unless these gaps are filled.

2.6 FUTURE RESEARCH DIRECTIONS IN IOT USING BLOCKCHAIN

Future research into IoT using blockchain focuses on several key areas, taking into account the peculiar challenges in this convergence of two technologies. Scalable blockchain architectures, together with layer-2 solutions and sharding, are probably the most important directions; these are intended to deal with huge amounts of data and transactions generated by IoT devices. IoT systems generally require energy-efficient consensus mechanisms, such as the PoS protocols, as many of these devices are resource-constrained and carry out low-power operations. Privileged-preserving technologies, such as zero-knowledge proofs and homomorphic encryption, would ensure the confidentiality of sensitive IoT data while continuing to protect user privacy. Overarching solutions for interoperability will be identified and developed for the creation of cross-chain protocols and standardised frameworks that allow frictionless interaction across blockchain platforms. Lastly, the integration of AI in IoT automation will enhance smart contracts and predict security threats in the blockchain-enabled IoT to improve performance with a security profile. These three research directions together go harmoniously as approaches towards advancing the integration of blockchain technology in IoT.

2.7 CONCLUSION

The integration of blockchain technology into the IoT system has enormous potential to transform many sectors, overcoming most of the challenges facing IoT, which relate to security, transparency, and data integrity. This chapter has examined the architecture of IoT utilising blockchain, its advantages, and key applications for a comprehensive understanding of how both technologies can be aligned to solve key concerns in data management and system efficiency. The combination of IoT and blockchain is very promising. Currently, however, it is surrounded by various challenges, such as scalability, storage management, high energy consumption, and also real-time processing. This chapter, in view of an extensive review of the literature and a gap analysis of the current research, points out areas that would allow further exploration and development. Accordingly, the storage problem is one of the significant bottlenecks of IoT systems because, by nature, blockchains are not natively designed for storing large heaps of data from IoT sources. Therefore, lightweight and efficient blockchains of scalable architecture, along with lightweight consensus algorithms and storage solutions, are mandatory for future research. The integration of IoT and blockchain promises to unlock innovations that could lead towards secure, scalable, and efficient systems that may benefit industries and applications across many categories.

REFERENCES

1. H. F. Atlam, A. Alenezi, M. O. Alassafi, and G. B. Wills, "Blockchain with the Internet of Things: Benefits, Challenges, and Future Directions," *International Journal of Intelligent Systems and Applications*, vol. 10, no. 6, pp. 40–48, Jun. 2018, doi: 10.5815/IJISA.2018.06.05.
2. K. R. Özyilmaz and A. Yurdakul, "Work-in-Progress: Integrating Low-Power IoT Devices to a Blockchain-Based Infrastructure," In *Proceedings of the 13th ACM International Conference on Embedded Software 2017 Companion, EMSOFT 2017*, Seoul, South Korea, Oct. 15–20, 2017, doi: 10.1145/3125503.3125628.
3. D. Puthal, N. Malik, S. P. Mohanty, E. Kougianos, and C. Yang, "The Blockchain as a Decentralised Security Framework [Future Directions]," *IEEE Consumer Electronics Magazine*, vol. 7, no. 2, pp. 18–21, Mar. 2018, doi: 10.1109/MCE.2017.2776459.
4. Y. Yu, Y. Li, J. Tian, and J. Liu, "Blockchain-Based Solutions to Security and Privacy Issues in the Internet of Things," *IEEE Wireless Communications*, vol. 25, no. 6, pp. 12–18, Dec. 2018, doi: 10.1109/MWC.2017.1800116.
5. N. Singh, H. P. Singh, A. Mishra, A. Khare, M. Swarnkar, and S. K. Almas, "Blockchain Cloud Computing: Comparative Study on DDoS, MITM and SQL Injection Attack," In *2024 IEEE International Conference on Big Data & Machine Learning (ICBDML)*, Bhopal, India, February 24–25, 2024, pp. 73–78, doi: 10.1109/ICBDML60909.2024.10577412.
6. P. P. Ray, D. Dash, K. Salah, and N. Kumar, "Blockchain for IoT-Based Healthcare: Background, Consensus, Platforms, and Use Cases," *IEEE Systems Journal*, vol. 15, no. 1, pp. 85–94, Mar. 2021, doi: 10.1109/JSYST.2020.2963840.
7. N. Adhikari and M. Ramkumar, "IoT and Blockchain Integration: Applications, Opportunities, and Challenges," *Network*, vol. 3, no. 1, pp. 115–141, Jan. 2023, doi: 10.3390/NETWORK3010006.
8. W. A. N. A. Al-Nbhany, A. T. Zahary, and A. A. Al-Shargabi, "Blockchain-IoT Healthcare Applications and Trends: A Review," *IEEE Access*, vol. 12, pp. 4178–4212, 2024, doi: 10.1109/ACCESS.2023.3349187.
9. N. Siegfried, T. Rosenthal, and A. Benlian, "Blockchain and the Industrial Internet of Things: A Requirement Taxonomy and Systematic Fit Analysis," *Journal of Enterprise Information Management*, vol. 35, no. 6, pp. 1454–1476, Nov. 2022, https://doi.org/10.1108/JEIM-06-2018-0140.
10. R. Roman, J. Zhou, and J. Lopez, "On the Features and Challenges of Security and Privacy in Distributed Internet of Things," *Computer Networks*, vol. 57, no. 10, pp. 2266–2279, Jul. 2013, doi: 10.1016/J.COMNET.2012.12.018.

11. S. Aich, S. Chakraborty, M. Sain, H. I. Lee, and H. C. Kim, "A Review on Benefits of IoT Integrated Blockchain-Based Supply Chain Management Implementations across Different Sectors with Case Study," In *International Conference on Advanced Communication Technology, ICACT*, vol. 2019, pp. 138–141, Apr. 2019, doi: 10.23919/ICACT.2019.8701910.
12. U. Majeed, L. U. Khan, I. Yaqoob, S. M. A. Kazmi, K. Salah, and C. S. Hong, "Blockchain for IoT-Based Smart Cities: Recent Advances, Requirements, and Future Challenges," *Journal of Network and Computer Applications*, vol. 181, p. 103007, May 2021, doi: 10.1016/J.JNCA.2021.103007.
13. P. Ratta, A. Kaur, S. Sharma, M. Shabaz, and G. Dhiman, "Application of Blockchain and Internet of Things in Healthcare and Medical Sector: Applications, Challenges, and Future Perspectives," *Journal of Food Quality*, vol. 2021, no. 1, p. 7608296, Jan. 2021, doi: 10.1155/2021/7608296.
14. A. Rejeb, et al., "Unleashing the Power of the Internet of Things and Blockchain: A Comprehensive Analysis and Future Directions," *Internet of Things and Cyber-Physical Systems*, vol. 4, pp. 1–18, Jan. 2024, doi: 10.1016/J.IOTCPS.2023.06.003.
15. M. A. Al-Shareeda, M. A. Saare, and S. Manickam, "The Blockchain Internet of Things: Review Opportunities, Challenges, and Recommendations," *Indonesian Journal of Electrical Engineering and Computer Science*, vol. 31, no. 3, pp. 1673–1683, 2023, doi: 10.11591/ijeecs.v31.i3.pp1673-1683.
16. H. F. Atlam, M. A. Azad, A. G. Alzahrani, and G. Wills, "A Review of Blockchain in Internet of Things and AI," *Big Data and Cognitive Computing*, vol. 4, no. 4, p. 28, Oct. 2020, doi: 10.3390/BDCC4040028.
17. S. K. Lo et al., "Analysis of Blockchain Solutions for IoT: A Systematic Literature Review," *IEEE Access*, vol. 7, pp. 58822–58835, 2019, doi: 10.1109/ACCESS.2019.2914675.

Section 2

Integration with Emerging Technologies

3 Understanding Blockchain Technology Adoption in Human Resource Management

A Technology Acceptance Model (TAM) Approach

Vishal Kumar, Sushil Kumar, Arvind Mewada, and Saaema Akhtar

3.1 INTRODUCTION

Blockchain technology, alongside other emerging technologies like artificial intelligence (AI), has significantly transformed various business sectors, including human resource management (HRM) [1]. Blockchain technology is known for decentralised and secure data management capabilities. To reshape processes, improve decision-making, and enhance operational efficiency, it has become integral to modern business operations [2]. Although most of the recent research has generally focused on newly adopted blockchain technologies in different areas, a limited number of studies have specifically looked at the potential of blockchain in HRM. Still, in spite of the parallelism of blockchain and AI, these two progressive technologies are advancing rapidly in this digital era [3]. AI refers to the capabilities through which computers can be taught to perform activities that a human would normally execute, and such a system can also learn from previous data and modify its tasks in accordance with current requirements. The genesis of the concept took place in the 1940s, but since then, it has undergone various stages of development [4]. Blockchain technology is already an innovation, and its progress is very high, especially in areas such as data storage security and managed decentralisation [5]. Better and more efficient decision-making and improvements in business processes have moved HR to be seen as a strategic partner of the organisations rather than simply a support function [6].

According to the use cases presented in this article, blockchain technology is especially useful in HRM workflows for talent management practices, such as recruitment and employee verification, as well as for dealing with sensitive information securely. It offers a trustworthy and transparent peer-to-peer network. As a result, the efficiency of HR activities increases, and incidents of fraud or data leakage are

DOI: 10.1201/9781003545620-5

prevented [7]. In addition, blockchain has also changed the way organisations manage HR functions in the employees' lifecycle, ranging from recruitment, onboarding, and performance appraisal to payroll processing [3,8]. Such tendencies are structured alongside the mounting prioritisation of HR practices of reconceptualising processes whereby blockchain is employed to enhance efficiency and simplify processes whilst increasing the accuracy of data.

While there is an enormous capability of blockchain technology application in HRM, there is still a need to increase the level of usage of the technology further. The lack of integration of blockchain into HRM practices is attributed to many organisations or institutions requiring support to do so due to the perceived complexity, security, and compatibility with the existing systems. Moreover, HR practitioners are key agents of organisational change who have a role in integrating technologies in their operational activities, which in this case is blockchain, but do not understand how it can be beneficial or easily implemented in their day-to-day office tasks [9]. Factors stemming from the HR practitioners who are relevant in blockchain adoption are important to utilise this technology to improve HRM and its processes. This research focuses on important factors influencing blockchain adoption in HRM, which are classified within the technology adoption model (TAM) framework. This study examines the views of HR practitioners about the use of blockchain technology and the factors that influence their choices to incorporate blockchain technology in HR systems [10]. In particular, this study seeks to address the following research questions:

1. Which factors significantly influence the acceptance of blockchain in HRM?
2. Which HRM tasks can Chinese professionals perform in a more efficient and useful manner?
3. What influences the levels of complexity, compatibility, security, and privacy in Blockchain acceptance?
4. What is the appropriate attitude towards the exchange of cryptocurrencies in organisations?

The purpose of this study is to achieve an in-depth appreciation of blockchain integration in HRM by addressing pertinent questions and offering a guideline on the prospects of successful adoption and implementation. It provides recommendations on how to address existing challenges to enhance HR functions, especially in the role of the HRM in the integration of blockchain technology. Billed as game-changers for HRM, blockchain technology improves speed, scalability, and secure data manipulation and may change the relationship of HR to practice in organisations. This chapter discusses some of the issues that affect the diffusion of blockchain and advises HR practitioners and organisations who wish to use blockchain on its advantages.

3.2 LITERATURE REVIEW

3.2.1 Adoption of Blockchain Technology in HRM

The HRM sphere has been revolutionised by blockchain in terms of security, transparency, and efficiency. The system integration has improved the delivery of services in HRM functions with the assurance of data consistency [11]. It is distributed and cannot be corrupted, which is very useful in HR for handling sensitive data (employee files, payment records, and documents pertaining to compliance) [12]. Blockchain provides a dependable form of record keeping and accounting, which is consistent with classical methods for maintaining information confidentiality and renders tamper-proof the changing of such information, thus increasing trust and accountability in various HR processes [13].

Because of blockchain, a range of HR activities, such as talent acquisition, contract management, and performance tracking, are impacted [14]. Blockchain reduces operational costs and positions HR as a strategic partner in achieving broader organisational goals by streamlining these processes [15]. As a result, the roles of HR have shifted from administrative support to a key player in driving organisational transformation through secure and efficient data management practices.

3.2.2 Technology Adoption Models in HRM

The adoption of blockchain in HRM has been examined through various theoretical models (the TAM and the technology-organisation environment) [16]. The TAMcomprises constructs such as Perceived Usefulness (PU) and Perceived Ease of Use (PEOU), which are widely used to understand individual and organisational acceptance of new technologies. It focuses on the behavioural aspects of technology adoption as users perceive the ease of use and usefulness of blockchain applications in HR. On the other hand, the technology-organisation environment model incorporates organisational and environmental factors such as top management support, technology readiness, and external influences [17]. Existing research studies that use these models have identified barriers and facilitators of blockchain integration within HR functions, such as perceived complexity, security concerns, and compatibility with existing systems [18]. Even though these insights are valuable afterwards, research gaps must be addressed to understand the factors that influence blockchain adoption in diverse organisational contexts.

3.2.3 Key Determinants: Compatibility, Complexity, Security, and Privacy

Blockchain technology adoption in HRM is influenced by several critical factors, which include compatibility, complexity, security, and privacy concerns. Compatibility means the seamless integration of blockchain with existing HR systems and processes. It indicates that high compatibility is linked to a smoother implementation process and greater acceptance among HR professionals, leading to higher PU of the technology [19]. However, complexity refers to how hard it is to perceive the adoption of blockchain technologies. In this respect, HR professionals find it too complex; in addition, the need for deep training serves as an obstacle to their adoption. Therefore, designing blockchain-based solutions that are user-friendly and easily integrable into daily workflows is critically important.

Moreover, security and privacy issues are paramount because HR data are sensitive. A decentralised structure, along with cryptographic features in blockchain, presents an important basis for effective data integrity and the reduction of the possibility of unauthorised access. Furthermore, addressing these critical points rightly is of vital importance to establish trust in blockchain-based

HR systems. By treating such key determinants, organisations can promote a healthy environment to adapt blockchain, thereby strengthening HR departments to apply this technology in their departments in a much more effective and secure way than before.

3.3 CONCEPTUAL FRAMEWORK AND HYPOTHESES DEVELOPMENT

3.3.1 Technology Adoption Model (TAM) Framework

TAM is the base of most models that are used to understand technology acceptance. Designing the model, Davis [20] originally came up with this model to predict whether a person will accept and use the technology or not. This model bases its argument upon the constructs of PU and PEOU. PU measures the extent to which an individual feels that application use would improve their job performance. Besides, PEOU is the extent to which a person feels that the use of the technology does not require much effort. TAM has been extensively applied in several organisational and institutional settings to explain technology adoption behaviour.

However, to capture the subtleties of blockchain adoption in HRM better, we extend this framework of TAM in this chapter by adding some additional constructs in the form of Relative Advantage, Compatibility, Complexity, and Security/Privacy concerns. The TAM with these extended constructs provides a much more holistic view of technology adoption from organisational and individual perspectives. This makes it suitable for a study on blockchain integration into HRM. It identifies the characteristics of blockchain technology because the decentralised technology is what makes it highly secure. It considers the influence of these attributes on the perceptions of HR professionals and, thus, on their adoption decisions [20].

3.3.2 Proposed Hypotheses and Conceptual Framework

It presents a conceptual model developed based on the eTAM framework, defining determinants of blockchain adoption in HRM. The variables considered for use in the proposed model include Relative Advantage, Compatibility, Complexity, Security/Privacy, and HR Readiness, which influence both PU and PEOU. Finally, this research develops hypotheses, whereby the constructs of PU and PEOU are supposed to influence the overall adoption of blockchain technology within an HR department.

A conceptual diagram for the proposed framework is presented in Figure 3.1, where arrows link through Relative Advantage, Compatibility, Complexity, Security/Privacy, HR Readiness, PEOU, and PU. These variables are applied to explain blockchain adoption in HRM. The arrow represents the hypothesised relationships and influence directions of these variables.

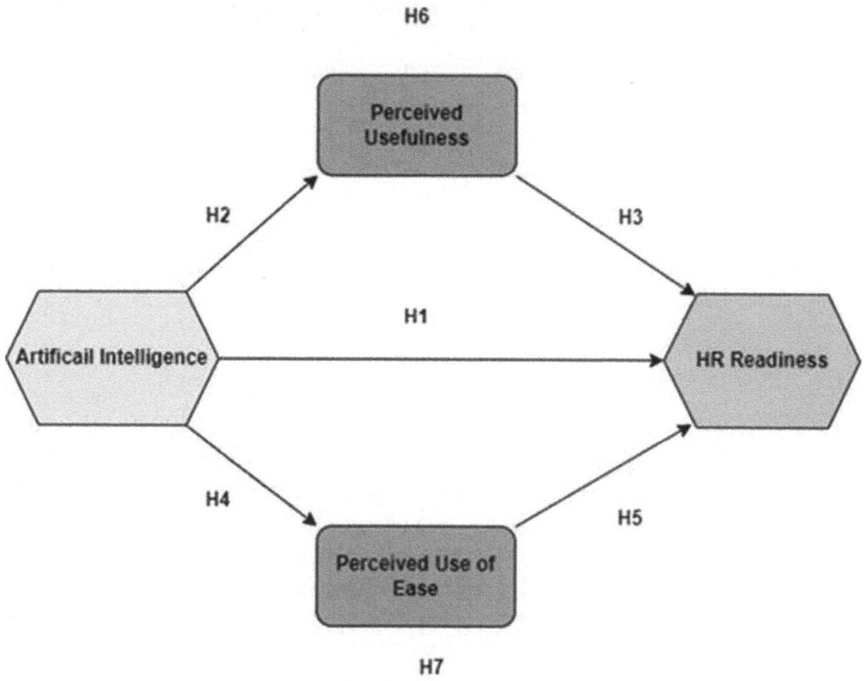

FIGURE 3.1 Conceptual framework for blockchain adoption in HRM.

3.3.3 Proposed Hypotheses

The following hypotheses are proposed for empirical testing:

1. **H1a:** Relative Advantage positively influences Perceived Usefulness (PU).
2. **H1b:** Relative Advantage positively influences Perceived Ease of Use (PEOU).
3. **H2a:** Compatibility positively influences Perceived Usefulness (PU).
4. **H2b:** Compatibility positively influences Perceived Ease of Use (PEOU).
5. **H3a:** Complexity negatively influences Perceived Usefulness (PU).
6. **H3b:** Complexity negatively influences Perceived Ease of Use (PEOU).
7. **H4a:** Security/Privacy concerns negatively influence Perceived Usefulness (PU).
8. **H4b:** Security/Privacy concerns negatively influence Perceived Ease of Use (PEOU).
9. **H5:** HR Readiness positively influences Perceived Usefulness (PU).
10. **H6:** Perceived Usefulness (PU) positively impacts the adoption of blockchain technology in HRM.

11. **H7a:** Perceived Ease of Use (PEOU) positively influences the adoption of blockchain technology in HRM.
12. **H7b:** Perceived Ease of Use (PEOU) positively influences Perceived Usefulness (PU).

This framework describes the technological and organisational contexts, meaning that it is applicable to HR professionals and organisations with an interest in implementing blockchain technology. This study intends to identify primary drivers and barriers towards the use of blockchain in HRM by testing these hypotheses and providing actionable insights for practitioners.

3.3.4 Practical Implications

Other variables in the extended TAM framework could specifically be relevant to blockchain. A richer understanding of the technology adoption process in HRM because it captures and sheds light on aspects such as organisational readiness, security/privacy concerns, and characteristic-type technology as shaping perceptions toward blockchain technology. It is germane to HR professionals and managers to use blockchain to better the HR functions, such as recruitment, payroll management, and employees' personal data security. However, the key factors discussed will help organisations prepare tailored strategies to eliminate adoption barriers and foster an ideal environment for efficient blockchain implementation in HRM. Thus, the current study contributes to the more general literature on technology adoption by providing insights into how HRM can change due to the integration of emerging technologies.

3.4 RESEARCH METHODOLOGY

3.4.1 Research Design

The research design defined the methodological approach taken in investigating the adoption of blockchain technology in HRM. This study based its theoretical framework on the TAM and took a two-level analysis, where both organisational and individual levels are considered. A quantitative method was applied to examine how the given factors for adoption, such as relative advantage, compatibility, complexity, security/privacy, and readiness, would impact the adoption of blockchain technology among HRMs. In other words, it is more focused on giving empirical evidence regarding the hypothesised determinants of blockchain adoption in HR practices by structured surveys and statistical analysis.

3.4.2 Sample and Data Collection

Data were collected using the online survey method as it is the most efficient and convenient method for easily reaching the largest number of participants. It targeted HR professionals from various verticals in India to understand their views and experiences regarding using blockchain technology. The sample population consists

TABLE 3.1
Demographic Characteristics of Respondents

Characteristic	Group	Frequency	Percentage
Gender	Male	98	81.6
	Female	22	18.4
Age Group	Young (<45 years old)	118	98.3
	Older (>45 years old)	2	1.7
Education	Bachelors	64	53.3
	Masters	42	35.0
	PhD	8	6.6
	Other	6	5.1
Profession	Employed	78	65.0
	Student	42	35.0
Work Experience	(>5 years)	84	70.0
	(5–10 years)	28	23.3
	(11–15 years)	2	1.6
	(16–20 years)	4	3.3
	(>20 years)	2	1.8

of 120 respondents who are HR managers, executives, and employees engaged in the technology-related functions of HR. Such a sample of diversification makes an understanding of the more general factors that influence the adoption of technology in different organisational contexts possible. Table 3.1 introduces the demographic characteristics of the respondents.

The sample consisted of 98 males (81.6%) and 22 females (18.4%). The sample was young, as 98.3% were less than 45 years old. Most participants held a bachelor's degree (53.3%) or a master's degree (35%), and 65% were employed in HR positions.

3.4.3 Questionnaire Development and Measurement Scales

A questionnaire was hence created to gather information regarding factors that influence the adoption of blockchain in HRM. The questionnaire consisted of 21 items that were formatted under various constructs based on the review of literature and framework of TAM. Every item was scaled to a rating of 5 points on a Likert Scale, where every respondent agreed or disagreed with the statements ranging between rating "1."

The importance of the Likert scale is based on the quantification of attitudes and perceptions, which is essential for analysis through mean and other statistical techniques [7]. It helps provide comparisons between different individuals and groups as well. However, there are two main limitations: the self-reporting system may be biased and it offers limited choices.

Table 3.2 portrays the conceptualisations of measurement items developed for each one of the key factors that contribute to the adoption of blockchain in HRM. Items are borrowed from relevant literature, ensuring constructs are empirically grounded.

TABLE 3.2
Factors Influencing Blockchain Adoption in HRM

ITEM	Ref.
Relative Advantage	
RA1: Blockchain in HR helps HR hire and retain key talent.	
RA2: Blockchain in HR helps reduce transactional work.	
RA3: Blockchain in HR makes HR functions more effective.	
Compatibility	
C1: Blockchain applications in HR are compatible with existing technology architecture.	[21]
C2: Blockchain applications are easy to integrate with existing HR functions.	[21]
C3: Customisation of blockchain applications in HR is easy.	[16]
Complexity	
CO1: Blockchain applications in HR are easy to use.	[21]
CO2: Blockchain applications are easy to integrate with existing HR workflows.	
CO3: Blockchain applications in HR are flexible to use.	
Security/Privacy	
SP1: Blockchain in HR has adequate security for employee data.	
SP2: Blockchain applications in HR protect employee confidentiality.	[17]
Human Resource Readiness	
HR1: HR professionals have the required skills to use blockchain in HR.	[18]
HR2: The HR department has the required technical resources to use blockchain.	
HR3: HR leaders in the organisation are keen to invest in blockchain.	
Perceived Ease of Use	
PEOU1: Adoption of blockchain in HR is easy.	[21]
PEOU2: It is easy to learn about blockchain in HR functions.	[20]
Perceived Usefulness	
PU1: With blockchain in HR, my role as an HR professional has become more effective.	
PU2: Blockchain in HR enhances the efficiency of HR functions.	
PU3: Organisations gain a strategic advantage by adopting blockchain in HR.	

3.5 DATA ANALYSIS AND RESULTS

3.5.1 Descriptive Statistics

The first step of data analysis is the point at which the descriptive statistics of the sample are used to provide a preliminary understanding of the composition and general characteristics of the dataset. It contains mean, SD, and frequency distribution summaries of the central tendency and variability of each variable in the study. Overall, distributional patterns help pinpoint any anomalies that might have an impact on later stages of analysis. Table 3.3 tabulates the distribution of important variables, which were measured based on a 5-point Likert scale. Relative Advantage holds the highest mean value at 4.30, which generally reflects the perception that this technology held more rewards for these respondents. However, Security/Privacy Concerns carried a somewhat lower mean value, which suggests user-based apprehension.

TABLE 3.3
Descriptive Statistics for Key Variables

Variable	Mean	SD	Minimum	Maximum
Perceived Usefulness (PU)	4.20	0.75	1.00	5.00
Perceived Ease of Use (PEOU)	4.05	0.82	1.00	5.00
Relative Advantage	4.30	0.70	2.00	5.00
Security/Privacy Concerns	3.95	0.65	1.00	5.00
HR Readiness	4.10	0.78	1.50	5.00

TABLE 3.4
Construct Reliability and Validity Metrics

Construct	CA	rho A	CR	AVE
Perceived Usefulness (PU)	0.80	0.78	0.88	0.68
Perceived Ease of Use (PEOU)	0.75	0.74	0.83	0.62
Relative Advantage	0.79	0.79	0.85	0.63
Security/Privacy Concerns	0.85	0.81	0.89	0.69
HR Readiness	0.77	0.73	0.84	0.61

3.5.2 Construct Reliability and Validity

Reliability and validity of constructs have been tested through the use of various statistical measures, such as Cronbach's alpha, rho A, composite reliability, and average variance extracted. From Table 3.4, it can be seen that all the constructs present high reliability and validity. The Cronbach's Alpha values range between 0.695 and 0.85, indicating good internal consistency. The highest value of 0.85 was observed for security/privacy construct. The values of rho A between 0.691 and 0.886 further confirmed the internal consistency of the constructs.

Furthermore, all of the constructs exceed the threshold of 0.70, which is the minimum composite reliability recommended, signifying good internal consistency. Average Varience Extraction (AVE) values ranged from 0.607 to 0.868 and so showed quite good convergent validity. Overall, it appears that the measurement model is also strong, and by extension, all the constructs.

Demonstrating high levels of reliability and validity, ensuring the accuracy and stability of the measurement items.

3.5.3 Structural Equation Modelling Outcomes

Structural equation modelling (SEM) was employed to examine the relationships between latent variables and to test the proposed hypotheses. The SEM analysis involved two key stages: evaluating the measurement model using confirmatory factor analysis and assessing the structural model to test the path coefficients and hypotheses.

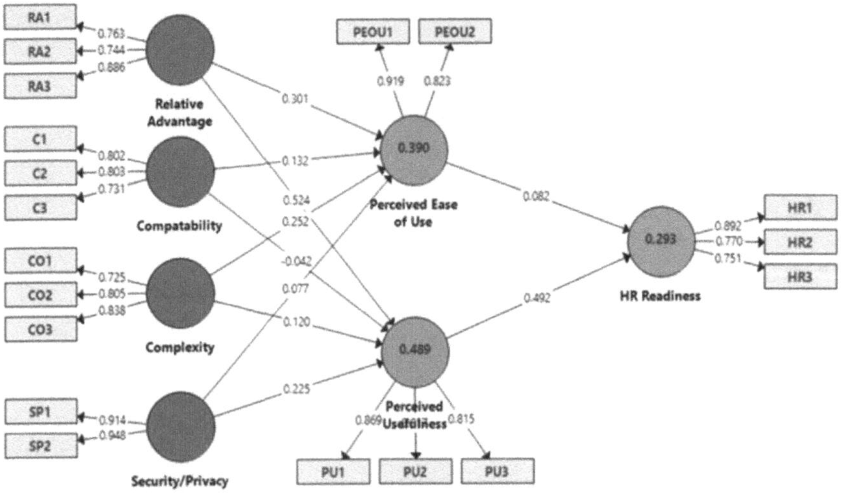

FIGURE 3.2 SEM path diagram.

TABLE 3.5
Path Coefficients and Hypothesis Testing

Hypothesis	PC	SE	*p*-value	Result
Relative Advantage → Perceived Usefulness	0.52	0.12	<0.01	S
Relative Advantage → Perceived Ease of Use	0.45	0.11	<0.01	S
Security/Privacy Concerns → Perceived Usefulness	0.34	0.14	<0.05	S
Complexity → Perceived Ease of Use	−0.15	0.10	0.10	NS

The SEM path diagram (see Figure 3.2) visualises the relationships between the constructs, indicating both significant and non-significant paths between the variables. The model fit was assessed using several indices, including the standardised root mean residual (SRMR), chi-square, and Normed Fit Index. The results indicated that both the saturated and estimated models exhibited suboptimal fit, with SRMR values exceeding the acceptable threshold, suggesting the need for model refinement. Specifically, the SRMR value was 0.09, above the recommended cutoff of 0.08. The chi-square value was 245.3; the p-value was less than 0.05, and the Normed Fit Index was 0.88, which did not reach a threshold of 0.90.

Several strong relationships were established through the PCs, SEs, and testing of the hypotheses, as shown in Table 3.5. For example, the PC for Relative Advantage and PU was approximately 0.52, with a p-value less than 0.01. This implies that there is a strong support of the hypothesis established. Similar in nature was the case of Relative Advantage and PEOU, for which this study found a PC of 0.45 with a p-value of less than 0.01. Results also showed that security/privacy concerns significantly affect PU along a path coefficient of 0.34 and a *p*-value of less than 0.05. These path coefficients measure the strength and significance of constructs' relationships; therefore, they point to important factors influencing the adoption of technology.

Table 3.5 provides a detailed breakdown of the hypotheses, PCs, and their statistical significance, highlighting the key determinants of technology adoption. The results underscore the importance of factors such as Relative Advantage and Security/Privacy Concerns in influencing both PU and PEOU. At the same time, Complexity did not show a significant impact on PEOU.

3.6 CONCLUSION

Blockchain technology has become pivotal in transforming HRM. While research on technology adoption in HR is extensive, studies on blockchain's specific applications in HRM are limited. An industry report outlined the benefits of blockchain in HR, but implementation details remain underexplored. Although various researchers have investigated blockchain deployment strategies and offered qualitative insights, there is a lack of comprehensive empirical studies using established models.

This study aimed to understand factors influencing blockchain adoption in HRM, focusing on HR professionals and key stakeholders. The research framework utilised the TAM, including organisational and individual-level analyses. The proposed model examined five categories: technological, organisational, environmental, "Perceived Usefulness," and "Perceived Ease of Use." Data were gathered through an online survey and analysed using partial least squares-structural equation modelling. This analysis provided a complete view of blockchain adoption in HRM, including assessments of reflective models, discriminant validity, and structural models.

REFERENCES

1. Moutaz Alazab, Salah Alhyari, Albara Awajan, and Ayman Bahjat Abdallah. Blockchain technology in supply chain management: an empirical study of the factors affecting user adoption or acceptance. *Cluster Computing*, 24(1):83–101, 2021.
2. Nagendra Singh, Harsh Pratap Singh, Anuprita Mishra, Anula Khare, Mamta Swarnkar, and Shekh Kulsum Almas, "Blockchain cloud computing: Comparative study on DDoS, MITM and SQL injection attack," In *2024 IEEE International Conference on Big Data & Machine Learning (ICBDML)*, Bhopal, India, February 24–25, 2024, pp. 73–78, doi: 10.1109/ICBDML60909.2024.10577412
3. Rupesh Kumar Dewang, Mahendra Pratap Yadav, Surbhit Awasthi, Om Raj, Arvind Mewada, and Kamlakant Laxman Bawankule. Data secure application: an application that allows developers to store user data securely using blockchain and IPFS. *Multimedia Tools and Applications*, 83(15):45491–45517, 2024.
4. Rukma Ramachandran, Vimal Babu, and Vijaya Prabhagar Murugesan. The role of blockchain technology in decision-making in human resource management: a review and future research agenda. *Business Process Management Journal*, 29(1):116–139, 2023.
5. Aradhana Sharma, Dhiraj Sharma, and Rajni Bansal. Challenges of adopting artificial technology in human resource management practices. In *The Adoption and Effect of Artificial Intelligence on Human Resources Management, Part B*, pp. 111–126, 2023. Emerald Publishing Limited.
6. Alpana Agarwal, Komal Kapoor, and Sandeep Walia. Modelling the barriers to blockchain implementation in human resource function. *International Journal of Quality and Reliability Management*, 41(8):2075–2094, 2024.
7. Heba Mohamed Adel and Raghda Abulsaoud Ahmed Younis. The interplay among blockchain technology adoption strategy, e-supply chain management diffusion, entrepreneurial orientation, and human resources information system in banking. *International Journal of Emerging Markets*, 18(10):3588–3615, 2023.

8. Vandana Madaan, Ram Singh, and Anil Dhawan. Use and applications of blockchain technology in human resource management functions. In *Effective AI, Blockchain, and e-Governance Applications for Knowledge Discovery and Management*, pp. 130–142, 2023. IGI Global.
9. Soumyadeb Chowdhury, Oscar Rodriguez-Espindola, Prasanta Dey, and Pawan Budhwar. Blockchain technology adoption for managing risks in operations and supply chain management: evidence from the UK. *Annals of Operations Research*, 327(1):539–574, 2023.
10. Mohini Damle, Prasanna Kulkarni, and Madhavi Damle. "Blockchain technology is used in talent retention and capability development in HRM," In *2023 5th International Conference on Inventive Research in Computing Applications (ICIRCA)*, pp. 1181–1188, 2023. Coimbatore, India.
11. Nagendra Singh and Yogendra Kumar. "Security and privacy requirements in 5G healthcare," In *Block Chain for 5G Healthcare Applications*, IET Publication, pp.1–30, Nov. 2021. https://doi.org/10.1049/PBHE035E_ch1.
12. Dongmei Yin. A human resource management model based on blockchain technology. *International Journal of Innovation and Sustainable Development*, 18(5–6):705–724, 2024.
13. Himani Mishra and M Venkatesan. Blockchain in human resource management of organisations: an empirical assessment to gauge HR and non-HR perspective. *Journal of Organizational Change Management*, 34(2):525–542, 2021.
14. Bharat Chillakuri and Prakash Attili. Role of blockchain in HR's response to the new normal. *International Journal of Organizational Analysis*, 30(6):1359–1378, 2022.
15. Thelma B Machado, Leonardo Ricciardi, and M Beatriz P. P. Oliveira. Blockchain technology for the management of food sciences research. *Trends in Food Science and Technology*, 102:261–270, 2020.
16. Chih-Hsien Lin and Wei-Hsiang Chen. A technology-organisation-environment (toe) framework based on scientometry for understanding the risk factors in sustainable water resources management. *Water Resources Management*, 37(15):5849–5869, 2023.
17. Nagendra Singh, S. P. Sasirekha, Amol Dhakne, B. V. Sai Thrinath, D. Ramya, R. Thiagarajan. IOT enabled hybrid model with loearning ability for E-health care systems. *Measurement: Sensors*, 24:1–14, 2022. https://doi.org/10.1016/j.measen.2022.100567.
18. .Deeba Kannan, Amutha Balakrishnan, K. Mekala Devi, Nagendra Singh, P. Angelin Kiruba, Ravindran Ramkumar, and Dhandapani Karthikeyan Hybrid rater to quantify and measure the severity of infection and spread of infection in muskmelon. *Bulletin of Electrical Engineering and Informatics (BEEI)*, 13(3):1667–1675, 2024. https://doi.org/10.11591/eei.v13i3.5432.
19. Marta Liesa-Oru´s, Cecilia Latorre-Cosculluela, Veronica Sierra-Sanchez, and Sandra Vazquez-Toledo. Links between ease of use, perceived usefulness, and attitudes towards technology in older people in university: a structural equation modelling approach. *Education and Information Technologies*, 28(3):2419–2436, 2023.
20. Afrizal Tahar, Hosam Alden Riyadh, Hafiez Sofyani, and Wahyu Eko Purnomo. Perceived ease of use, perceived usefulness, perceived security, and intention to use e-filing: the role of technology readiness. *The Journal of Asian Finance, Economics and Business*, 7(9):537–547, 2020.
21. Mohd. Aquib Ansari and Dushyant Kumar Singh. Human detection techniques for real time surveillance: a comprehensive survey. *Multimedia Tools and Applications*, 80:8759–8808, 2021. https://doi.org/10.1007/s11042-020-10103-4.

4 Blockchain Technology's Key Components, Security Challenges, and Its Evolving Role in Modern Digital Infrastructure

N. Sharmila Banu, K. Deepa, and Mastan Vali Shaik

4.1 INTRODUCTION

Blockchain technology was introduced in 2008, and the distributed database was used to track Bitcoin transactions. The successful mining of the genesis block by Nakamoto in 2009 validated the concept of blockchain [1]. The framework concept included the electronic currency system, which is based on a peer-to-peer network, blockchain technology, encoding and timestamps [2]. Peers can successfully transfer values through transactions, negating the necessity for central control to safeguard client privacy and stop identity intruders [3]. From its initiation, from the first crypto-currency to the present blockchain-related approaches for industry 5.0, blockchain technology has been implemented into the communications of several organizations that require precision, probity, and consistency [4–6].

Nevertheless, the widespread adoption and continuous advancement of new technologies introduce inherent risks and raise concerns regarding blockchain technology. A smart contract is a code snippet made publicly accessible on the Ethereum network [7]. The bitcoin cryptocurrency sustained to gain significance and reputation in 2013, despite Satoshi Nakamoto's disappearance from medium, code contributions, and papers in 2011. The primary factor contributing to this increase in popularity was the currency's acceptance by websites and fund-based startups. The technology underpinning Bitcoin is constantly evolving, and human errors and varying standards can influence its implementation. Cryptocurrencies represent the pioneering and most widely acknowledged use of blockchain technology, with Bitcoin standing out as the most prominent instance. They leverage blockchain as a decentralized public ledger to document all transactions, successfully addressing the double spending issue

DOI: 10.1201/9781003545620-6

linked to digital currency. This framework enables secure and transparent transactions, eliminating the necessity for a trusted intermediary [8].

Rather, there is a blockchain database accessible via the dispersed computer network. Each computer connected to the network is referred to as a node, and each node is given a duplicate copy of the blockchain, which is downloaded automatically. Digital signatures for transactions are created using public key cryptography, which employs two keys, of which there are two types: public and private. There is a mathematical relationship between these two keys. The transactions are more difficult to crack because it is nearly impossible to predict these keys due to the intricacy of the math involved.

A message to be delivered is signed and encrypted using the public key, which may be decrypted by the intended receiver using their private key. Data about every new transaction is distributed to every node in the blockchain database to keep it updated and functioning as a "Worldwide Ledger." Blockchain technology has been utilized in numerous sectors owing to its distinctive characteristics, including decentralization, security, and transparency. In the healthcare sector, it can enhance patient identity management by offering a reliable and standardized identity for every individual. Its application in information systems facilitates secure data sharing and storage.

Furthermore, incorporating blockchain into wireless networks can bolster security and data integrity, proving an essential resource for managing network assets and user information. In addition, blockchain delivers high-security standards for Internet of Things applications, protecting against unauthorized data access [9,10]. Blockchain technology has considerable potential for transforming electronic health records by improving security, facilitating interoperability, and empowering patients, all while tackling several challenges within the healthcare system [11].

4.2 COMPONENTS OF BLOCKCHAIN TECHNOLOGY

4.2.1 Decentralization

Decentralization is the progression by which responsibilities and abilities are transmitted from a unified control to each participating body. Blockchain technology prevents the possibility of centralized control, even though each blockchain user or miner accepts a copy of the transaction database. When the miners validate the transaction, a new block is added. The network functions on a peer-to-peer basis in a decentralized setting. This blockchain feature is one of the key components researchers use to construct the Ethereum virtual currency.

4.2.2 Consensus Building

The integrity of data stored on the blockchain is upheld, in part, by the consensus model. However, when the consensus system malfunctions, a range of issues can emerge, such as blockchain connectivity problems, identity failures, authority discrepancies, node verification challenges, and inadequate network representation [12,13]. The efficiency and applicability of a consensus protocol depend on three key characteristics: security, activeness, and fault tolerance. Security ensures that the protocol is both compatible and safe, with every node generating outcomes that align with the protocol's guidelines and are legitimate. Activeness requires that non-value nodes remain operational to ensure that the consensus process can generate a value. Lastly, fault tolerance allows the protocol to recover even if a node involved in the consensus process ceases to function. These characteristics are crucial for maintaining the stability and reliability of blockchain systems.

4.2.3 Transparent

Ten-minute intervals of transactions are adapted by the blockchain network, which periodically self-audits the ecology of a digital asset by checking in with itself every 10 minutes. The grouping of all these communications is termed a block. Precision and the impossibility of corruption are the two qualities that emerge. The utilization of blockchain technology has the potential to enhance the security, efficiency, and transparency of digital transactions. Such advancements can motivate economic growth and promote the greatest trust in digital systems [14].

4.2.4 Open Source

Users of closed-source, decentralized applications must have faith that the data are not accessible from a central source and that the program is decentralized. Closed-source applications prevent people from adopting them. When an application is meant to collect, keep, or transfer user dues, operating on a closed-source network is obviously disgusting. Consumers prefer open-source participation even if it is feasible to develop a closed-source decentralized program. This is because the complexity involved in achieving the expected outcome would be fatal. An open-sourced, decentralized application modifies based on how corporate operations are organized, which was previously dependent on the net as a unifying factor.

4.2.5 Identity and Access

Three primary factors determine a blockchain's uniqueness and convenience: grouping, private, public and both. Users are not allowed to validate block transactions or write digital agreements on a private blockchain. The classic business and governance concepts are suitable [15]. Public blockchains are made to eliminate intermediaries from transactions while maintaining security. Anyone with internet access can join a public blockchain network by generating smart contracts and participating in block verifications. Grouping blockchains enables a small number of predetermined nodes to have total control and are partially private. It serves as a stand-in for letting any arbitrary person with an internet connection confirm transactions. Platforms such as private blockchains offer transaction efficiency and secrecy. The selection of a blockchain platform is important in digital settings as Bitcoin, Ethereum, and Hyperledger Fabric each provide distinct advantages that address various requirements related to data integrity and security [14].

4.2.6 Autonomy

The primary goal of blockchain technology is to transfer confidence from individual, centralized control to the entire group, which lacks any intervention. Information can be safely updated and sent between any nodes in the blockchain system. An association that abides by a group of predetermined systems approved and executes when the requirements are met is known as a smart contract; it is referred to as a decentralized autonomous organization, which is also known as a decentralized autonomous corporation. In blockchain, blocks accumulate the transaction record and the specifics of the smart contract.

4.2.7 Immutability

It is described as data that does not change throughout time. In the context of blockchain, unchanged data with regard to information contained in the block is crucial. This framework guarantees that once information is entered into a block, it cannot be modified without altering all subsequent blocks. This immutability characteristic is a fundamental aspect of blockchain technology, ensuring the integrity and reliability of the system [16]. The recipient of the data is confident in its accuracy and integrity. The immutable nature of blockchain is particularly beneficial for databases involved in financial transactions, as entries are preserved and cannot be altered unless an individual gains control over the nodes of the network.

4.2.8 Anonymity

This part only needs a miner's blockchain address; no additional information is needed; hence, anonymity helps to resolve trust concerns. This concept utilizes a group of entities that is impossible to differentiate. The sender and recipient sets are the two categories into which the confidentiality set can be separated among them. The components of blockchain technology are illustrated in Figure 4.1.

4.3 WORKING PROCESS OF BLOCKCHAIN

Blockchain is a publicly accessible ledger that operates through several key processes, outlined below and illustrated in Figure 4.2. First, a node or user broadcasts data to the network, initiating the transaction. Next, the receiving node verifies that the data transmitted over the network is authentic. Once validated, the data are saved to a block. The block is then subjected to proof of work, a process that requires validation by all nodes or clients within the network to ensure the authenticity of the operation. Finally, a consensus mechanism is applied to the validated data, after which the block is added to the blockchain.

4.4 ISSUES AND CHALLENGES IN BLOCKCHAIN TECHNOLOGY

Conquering all the challenges and tackling the issues affecting the contemporary world takes time. The blockchain is used in this situation. It will take time to overcome all the hurdles despite many possibilities. Adopting blockchain is very

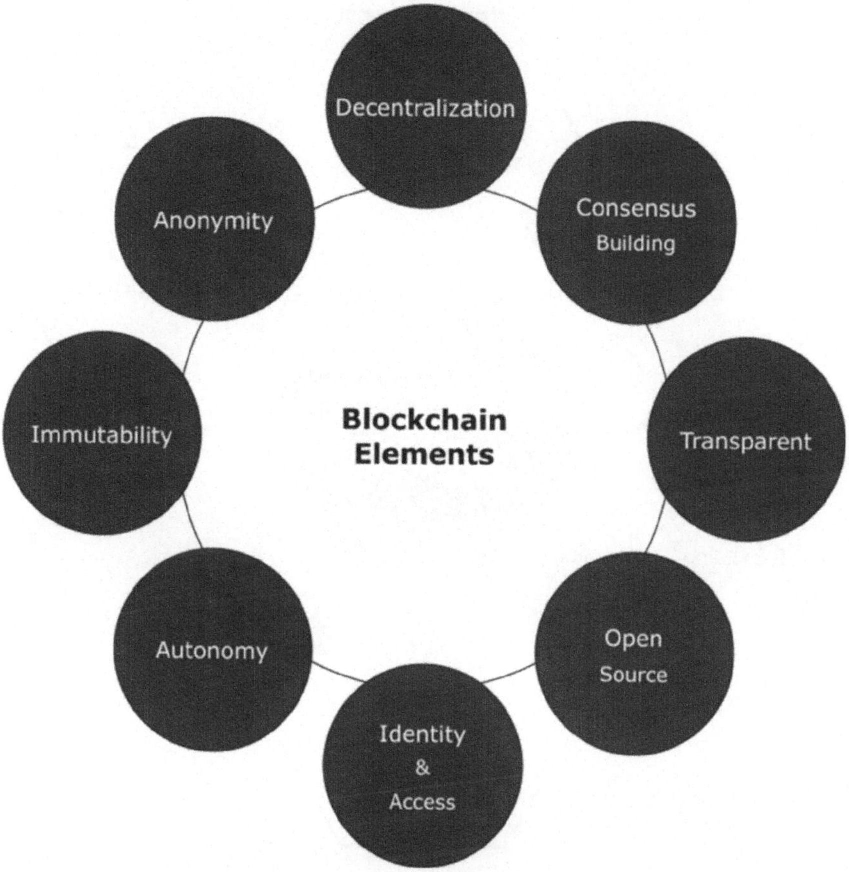

FIGURE 4.1 Components of blockchain technology.

effective, but there is a chance of facing a lot of complexity that influences developers and users. The issues and challenges in blockchain technology are described with the following points:

4.4.1 Scalability

Although transactions per second are a popular measure of blockchain execution scalability, Transaction Processing System (TPS) refers to the number of computations per second because transaction costs and complexity can fluctuate. A network can handle more computations at any given time as more transactions pass [17]. The primary challenge in scaling the execution layer is achieving higher computation rates per second without significantly raising the hardware specifications on each complete node that verifies a block of transactions. As transaction volumes rise, the ability of the blockchain to process these transactions effectively may become a restrictive factor. This concern is especially pertinent for startups that could undergo swift expansion and need a system capable of managing increased demand while maintaining optimal performance [18].

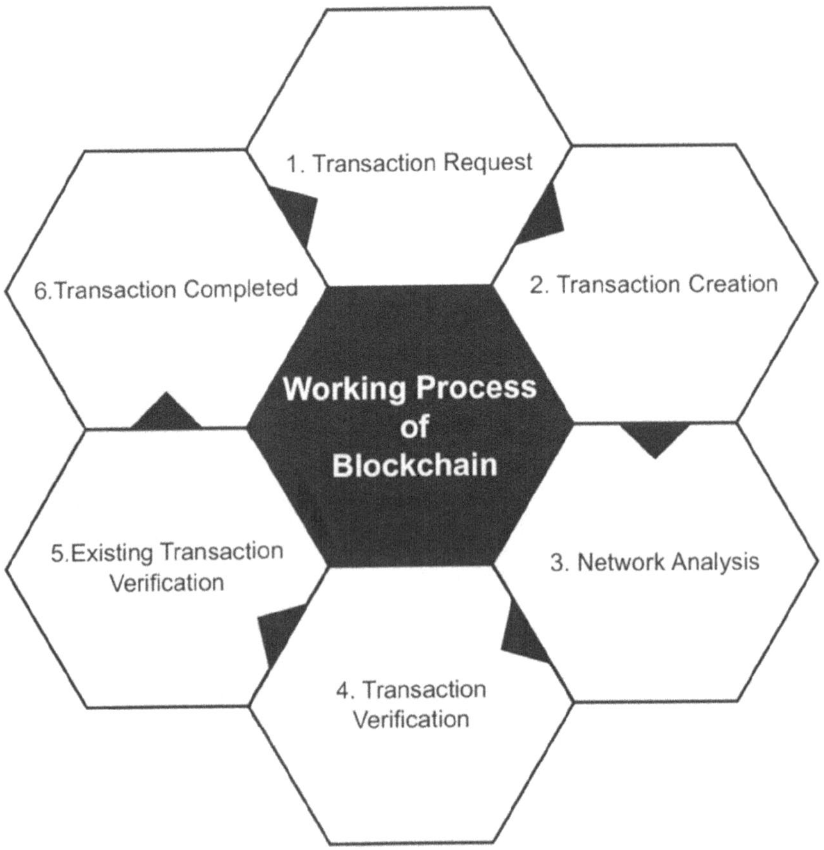

FIGURE 4.2 Working process of blockchain.

4.4.2 Data Protection

The major complication to extensive acceptance is the lack of assurance among blockchain service contributors. There are two aspects related to this concern: organizations might not have confidence based on the security provided by the technology and other users in a network. Every transaction is thought to be secure, confidential, and validated. This is accurate despite the lack of control over approving and authenticating the transactions due to the decentralized nature of the network. An essential element of every blockchain network is the consensus approaches that assist universal concurrence over the present situation of the distributed database across the whole network. Business leaders have discovered that users are more confident in private blockchains without recognized providers. Blockchain introduces the danger of privacy leaking and guarantees open and transparent transactions. Using techniques such as clustering analysis, attackers can successfully deduce the genuine identity of blockchain users.

4.4.3 Privacy

Blockchain was intended to be shared with the community. This implies that everyone can view the information added to the blockchain. The supplementary particulars of a transaction are visible, even though the data are unknown when blockchain directory addresses are used as identifiers. Private blockchains limit who can interpret transactions and participate as a node. Though it has weaknesses, a private blockchain is a method for a corporation to use blockchain technology without worrying about disclosing any information to the public.

A private blockchain is not fully decentralized because it contains a governing body that decides the participation of the nodes. This may cause people to have less faith in a blockchain-based result. The transparency of public blockchains can result in privacy challenges, as sensitive information might be accessible to all network participants. This creates significant risks for sectors that manage confidential data, including healthcare and finance [19].

4.4.4 Selfish Mining

Selfish mining is a hidden method used by a few users of distributed digital database systems or blockchain networks to commit fraud. They accomplish this by covertly mining and preserving a block. These miners then reveal their hidden chains to the public when they are longer than the actual ones, and the blockchain system chooses the longest chain as the genuine chain. Blockchain technology provides potential solutions for enhancing data security; however, it also introduces various security challenges that must be tackled to guarantee its successful application in healthcare and other industries [20]. Few security attacks affect the transaction, and the user's confidence is as follows:

1. **Mitigating 51% Attack Risks:** The main function of miners is to validate the transaction request and data so they can investigate the following block even more. Because a 51% attack has the potential to change the entire blockchain, it is perhaps the riskiest in the blockchain sector. This kind of attack is more likely to occur in the chain when there are not enough miners. To prevent 51% assaults, raise the hash rate, enhance mining pool supervision, and avoid providing proof for the task [20].
2. **Sybil Attack:** In this attack, hackers fabricate bogus network nodes and overwhelm the indented network with fictitious identities, causing a system meltdown and disturbing chain transactions. The proper utilization of consensus techniques should be considered to prevent the attack. Keep track of all the other nodes and concentrate on the nodes that are only forwarding blocks from the individual to prevent Sybil attacks.
3. **Phishing Attacks:** This type of attack is widely increasing nowadays and resulting in actual problems. The main aim of a phishing attack is to hack the user's credentials by the intruders. They are able to send the administrator emails that appear authentic and collect the qualification details and other related data. At the same time, the user submits their login information

through a joined bogus hyperlink. Installing trustworthy anti-malware software or malicious link detection software and maintaining systems and software updates will improve the security of the browser and device. Ignore clicking on unidentified links and turning on Wi-Fi while doing online transactions to guard against phishing attempts, especially when utilizing an electronic wallet or other sensitive data.

4. **Routing Attacks:** This is a major security issue that should be considered seriously, followed by privacy issues. Hackers can utilize the identity of an account to seize information submitted to internet service providers. There is a chance for intruders to hack the sensitive data or assets without the knowledge of the owners. Users should ensure safe routing mechanisms are implemented with digital signatures, encode owner data, use strong passwords, modify them frequently, and tutor themselves regarding the hazards associated with data security to avoid outing attacks.
5. **Private Key Security Attacks:** Blockchain technology is based on the concept of asymmetric key cryptography. Security issues arise when the key is not utilized properly, and the implementation process is not done perfectly. There is a chance of acquiring the private key from the user's public key.
6. **Blockchain Endpoint Vulnerabilities:** The sensitivity of blockchain endpoints is one of the serious concerns in blockchain technology, where hackers can obtain the users' passwords, keep track of their activities, and aim at the devices. To be safe, the user should regularly monitor the system, time, and location and change the password by providing a strong password and access to the device.

4.5 CONCLUSION

Blockchain technology is extremely accepted and reviewed due to its disseminated communications and end-to-end quality. This distinctiveness has the potential to maintain various criteria in numerous contexts and uses. With the rapid development and improvement of blockchain, we believe that it will rapidly turn into a very frequent and recognized circumstance. Blockchain can be compared with the internet a few decades ago to a certain extent. Since the core of blockchains is secure and supportive, progressively, the transactions that require security and non-redundancy will consider this expertise. Blockchain technology is expected to become the standard as it matures, even though certain limits currently make many creative applications challenging to implement.

REFERENCES

1. Dewang, R. K., Yadav, M. P., Awasthi, S., Raj, O. Mewada, A., and Bawankule, K. L. Data secure application: an application that allows developers to store user data securely using blockchain and IPFS. *Multimedia Tools and Applications*, *83*(15), 45491–45517, 2024.
2. A. Sobecki, S. Barański, and J. Szymański, J. "Privacy-preserving, scalable blockchain-based solution for monitoring industrial infrastructure in the near real-time", *Applied Sciences*, *12*(14), 7143, 2022.

3. Singh, N., Singh, H. P., Mishra, A., Khare, A., Swarnkar, M., and Almas, S. K., "Blockchain cloud computing: comparative study on DDoS, MITM and SQL injection attack," In *2024 IEEE International Conference on Big Data & Machine Learning (ICBDML)*, Bhopal, India, 2024, pp. 73–78. https://doi.org/10.1109/ICBDML60909.2024.10577412
4. Bodkhe, U., Tanwar, S., Parekh, K., Khanpara, P., Tyagi, S., Kumar, N., and Alazab, M. "Blockchain for Industry 4.0: a comprehensive review", *IEEE Access*, *8*, 79764–79800, 2020.
5. Khan, S. N., Loukil, F., Ghedira-Guegan, C., Benkhelifa, E., and Bani-Hani, A. "Blockchain smart contracts: applications, challenges, and future trends", *Peer-to-Peer Networking and Applications*, *14*, 2901–2925, 2021.
6. Rupa, C., Midhunchakkaravarthy, D., Hasan, M. K., Alhumyani, H., and Saeed, R. A. "Industry 5.0: ethereum blockchain technology-based DApp smart contract", *Mathematical Biosciences and Engineering*, *18*(5), 7010–7027, 2021.
7. Jafar, U., Aziz, M. J. A., and Shukur, Z. "Blockchain for electronic voting system—review and open research challenges", *Sensors*, *21*(17), 5874, 2021.
8. Karthikeyan, D., Amutha, B., Mekata Devi, K., Singh, N., and Anglin Kuruba, P. "Hybrid rater to quantify and measure the severity of infection and spread of infection in muskmelon", *Bulletin of Electrical Engineering and Informatics (BEEI)*, *13*(3), 1667–1675, 2024. https://doi.org/10.11591/eei.v13i3.5432
9. Krichen, M., Ammi, M., Mihoub, A., and Almutiq, M. "Blockchain for modern applications: a survey", *Sensors*, *22*(14), 5274, 2022.
10. Singh, S. P., Ansari, M. A., and Kumar, L. "Analysis of website in web data mining using web log expert tool", In *2023 IEEE 12th International Conference on Communication Systems and Network Technologies (CSNT)*. IEEE, 2023. Bhopal, India.
11. Goenkaa, N., Sharmab, A. K., Tiwaric, S., Singhd, N., Yadave, V., and Chadaga, S. P. K. "A regularized volumetric ConvNet based Alzheimer detection using T1-weighted MRI images", *Cogent Engineering*, *11*(1), 2024. https://doi.org/10.1080/23311916.2024.2314872
12. Singh, N., and Kumar, Y. "Security and privacy requirements in 5G healthcare", In *Block Chain for 5G Healthcare Applications*, IET Publication, pp.1–30, 2021. https://doi.org/10.1049/PBHE035E_ch1
13. Benčić, F. M., and Žarko, I. P. "Distributed ledger technology: blockchain compared to directed acyclic graph", In *2018 IEEE 38th International Conference on Distributed Computing Systems (ICDCS)*, pp. 1569–1570, 2018. Vienna, Austria.
14. Maariz, A., Wiputra, M. A., and Armanto, M. R. D. "Blockchain technology: revolutionizing data integrity and security in digital environments", *International Transactions on Education Technology (ITEE)*, *2*(2), 92–98, 2024.
15. Bayan, T., and Banach, R. "Exploring the privacy concerns in permissionless blockchain networks and potential solutions", In *2023 IEEE International Conference on Smart Information Systems and Technologies (SIST)*, pp. 567–572, 2023. Astana, Kazakhstan.
16. Habib, G., Sharma, S., Ibrahim, S., Ahmad, I., Qureshi, S., and Ishfaq, M. "Blockchain technology: benefits, challenges, applications, and integration of blockchain technology with cloud computing", *Future Internet*, *14*(11), 341, 2022.
17. Singh, N., Sasirekha, S. P., Dhakne, A., Sai Thrinath, B. V., Ramya, D., and Thiagarajan, R. "IOT enabled hybrid model with learning ability for E-health care systems", *Measurement: Sensors*, *24*, 1–14, 2022. https://doi.org/10.1016/j.measen.2022.100567
18. Putro, A. N. S., Mokodenseho, S., Hunawa, N. A., Mokoginta, M., and Marjoni, E. R. M. "Enhancing security and reliability of information systems through blockchain technology: a case study on impacts and potential", *West Science Information System and Technology*, *1*(1), 35–43, 2023.

19. Idrees, S. M., Nowostawski, M., Jameel, R., and Mourya, A. K. "Security aspects of blockchain technology intended for industrial applications", *Electronics*, *10*(8), 951, 2021.
20. Wenhua, Z., Qamar, F., Abdali, T. A. N., Hassan, R., Jafri, S. T. A., and Nguyen, Q. N. "Blockchain technology: security issues, healthcare applications, challenges, and future trends", *Electronics*, *12*(3), 546, 2023.

5 Security Mechanisms and Challenges in Blockchain Technology

A Comprehensive Framework for Addressing Threats and Enhancing Integrity

J. S. Prasath and Mohd Sameen Chishti

5.1 INTRODUCTION

Blockchain technology has become more popular and rapidly growing in recent years. Blockchain networks and blockchain technology are used in various industries, including cyber-security, information and communication technology, finance and banking, energy and infrastructure, entertainment, transportation, supply chain management, insurance, and healthcare technology. Blockchain technology gathers and stores information in groupings known as blocks, and each block can be used to grasp a set abstraction of data. The hybrid privacy-preserving access control scheme is proposed for blockchain with confidentiality considerations [1]. A prototype system confirms its feasibility and effectiveness, achieving low network latency, efficient encryption, and robust data security. The scope, concerns, and potential solutions related to the segregation of blockchain technology and the IoT are addressed [2]. A secure data management scheme that utilizes encryption, integrity verification, and an integrated communication network has been developed [3]. The security and privacy of blockchain are enhanced by fully homomorphic encryption. The blockchain-based Intrusion Detection System is proposed to enhance the Internet of Things (IoT) network security [4]. The machine learning algorithms detect Distributed Denial of Service (DDoS) attacks. The protocol to ensure security over the permissioned blockchains is presented [5].

The blockchain mechanism is utilized to maintain the stored information and improve the security level of the protocol. Blockchain is a type of database technology configured to provide a secure and unalterable record of transactions. A secure blockchain-enabled vehicle-to-vehicle (V2V) communication system is projected to

DOI: 10.1201/9781003545620-7

improve the security of vehicles and secure data sharing and communication among vehicles [6]. These algorithms can certify users, identify illegitimate access, and constitute secure vehicle communication. When a block is filled with data, it is chained to the previous full block, producing a data chain. The data is organized into blocks, each containing a packet of transactions. A novel security mechanism based on ledger technology is proposed [7].

Blockchain technology operates on both public and private networks. Any party may gather the information on public networks, which are internet-based communications, to authorize transactions. The privacy mechanism is achieved through machine learning [8]. Private blockchain networks allow only the authorized enterprise to access the data. Organizations should form the members and provide rights and permissions to access the information related to the transactions. A novel flexible network infrastructure for smart cities is formulated that integrates software-defined networks and blockchain technology [9].

The software-defined networks will identify the possible attacks and send the secured data to the blockchain. A blockchain operates on a distributed network of nodes. This makes it incredibly difficult for intruders who want to purposely collapse the network or manipulate the essential information of an organization. The problems related to real-time big data storage and security concerns are addressed [10].

The blockchain secures and validates transactions using intelligent mathematical algorithms. It is impossible to alter the data since each transaction is encrypted and connected to the previous one. The security in cloud computing is accomplished through an authentication mechanism [11]. The major security goals are confidentiality, integrity (ensuring the accuracy and reliability of information), and availability (making data accessible to authorized parties). The complexity is to achieve a higher level by considering several security goals. The authorization mechanism is proposed based on an association blockchain for internet-enabled equipment [12].

Blockchain technology is specially designed to achieve data integrity. Blockchain technology provides a guaranteed mode to track the motion of products between two places, which might improve supply chain management. A regulated blockchain and a secure and automated key refreshment scheme for IoT devices are proposed [13].

This proposed scheme permits users to validate the originality of the security keys and launch trust in an IoT system. The key management scheme is proposed based on blockchain for usage in internet-enabled devices [14]. The challenges and trends towards blockchain security are proposed [15]. The risk analysis, risk categories, and real-time attacks of blockchain are addressed. The blockchain-based information sharing between devices is discussed [16]. It is designed for usage on internet-enabled devices.

A public blockchain network does not require authentication, allowing everyone to access the information. All blockchain users have identical rights to read, edit, and authorize the blockchain. The participants can access and validate the transactions being processed on the public blockchain. The blockchain mechanism is proposed for internet-enabled equipment [17]. This proposed scheme permits the storage of copious amounts of information. The issues in secure communication of patient information are addressed [18]. The blockchain-based authentication mechanism is projected to meet the security challenges [19].

Improvements in various security schemes based on blockchain are proposed [20]. The security and privacy-preserving mechanism is proposed based on blockchain technology [21]. The concept of blockchain and its related components are addressed [22]. The solutions for improving blockchain security are also discussed. The blockchain mechanism for intelligent internet-enabled devices is proposed [23].

The solution for protecting the information is addressed. Private blockchains, also called managed blockchains, can be controlled by a single organization. Private blockchains do not provide access to all users, and only authorized parties are allowed to access the networks. The various problems in securing the internet-enabled blockchain technology are discussed [24]. The combination of the internet and blockchain technology is analysed, along with the research issues and trends in blockchain applications [25]. The security challenges, vulnerabilities, attacks, and potential research of smart contracts in IoT are addressed [26]. This work also discusses the potential applications of smart contracts. The blockchain-based technique for identifying the illegitimate node is addressed [27].

The contributions of blockchain for managing wireless sensor network security are discussed. Hybrid blockchains incorporate data from both private and public connections. Industries can set up private, permission-based systems along with public networks. The authorized parties only access the confidential information, whereas the rest of the information is made public access. Hybrid blockchains can provide public access to digital currency while keeping bank-owned currency private. The problems involved in securing Bitcoin are addressed [28].

Every device in a network is enabled to vote for the slave device through a consensus scheme [29]. The architecture of blockchains, security concerns, and certainty problems are addressed [30]. Blockchain can act against data tampering and allows a larger amount of data to be shared among networks. Consortium blockchains are widely used in banking, finance, logistics, healthcare, and insurance. The data communication security for internet-enabled devices is proposed [31]. The blockchain-based security framework is proposed [32].

This security mechanism prevents attacks during data communication. The core security challenges of blockchain technology include vulnerabilities of smart contracts, Sybil attacks, management of private keys, internal attacks, and DDoS attacks. Sybil attacks can be identified and prevented by incorporating reputation-based systems and proof-of-stake schemes. Private key management can be attained through multi-factor authentication, providing an additional security layer. Internal attacks in blockchain can be overcome by utilizing strong encryption methods such as AES-256 (Advanced Encryption Standard), secure routing protocols, strengthening access control mechanisms, and monitoring malicious activity. DDoS attacks in blockchain can be rectified by creating secure smart contracts, incorporating application firewalls, and monitoring the network traffic.

5.2 SECURITY CHALLENGES IN BLOCKCHAIN

The major challenge in the design of blockchain is to accomplish data integrity. The cyber security threats associated with public blockchain are confidentiality and availability. Data integrity can have its threats associated with the source and quality of the registered data. The challenges in securing blockchain technologies are lack of regulation, Interoperability, and improper management of private keys.

5.2.1 Lack of Regulation

The growth towards legislation is continuing, as blockchain technology is still relatively new. The lack of regulations in blockchain transactions makes it easier for intruders to utilize weaknesses without issues. The lack of regulation issue can be rectified with the support of blockchain creators, enterprise leaders, and officials. They can serve jointly to make a stable regulation that protects users while promoting innovation. It is essential to create standards and best practices among Blockchain consumers.

5.2.2 Interoperability

The different standards and protocols utilized by blockchain technology. The possibility of security threats occurs during the transmission of data across various blockchains due to the limitations of interoperability. Interoperability issues can be overcome by creating flexible blockchain solutions. The cross-chain bridge mechanism can be utilized to solve the interoperability problem. This cross-chain bridge allows blockchain networks to interact with isolated blockchains to assist data communication.

5.2.3 Poor Management of Private Keys

The private keys must be kept confidential in blockchain technology to protect against unauthorized access to their digital assets. Most users are unaware of the best methods for managing private keys and could intentionally disclose their keys to intruders. The solution for effectively managing private keys is to utilize biometric-based digital signatures. The consumer's biometric details are used to generate a secret key every time and directly remove it from memory. The best practices that need to be followed for private key management are utilizing strong cryptographic algorithms, secure key generation mechanisms, and storing encrypted keys in separate cryptographic modules.

5.3 SECURITY THREATS IN BLOCKCHAIN

Blockchain is related to Bitcoin and cryptocurrencies, and it is widely used in manufacturing sectors, supply chain management, mining, agriculture, energy, and healthcare. Cryptography is used to secure transactions across blockchain networks. The security against cyber-attacks is difficult because of decentralization. The various security threats in blockchain networks are shown in Figure 5.1. The various security threats of blockchain networks are mining threats, network threats and smart contract threats.

5.3.1 Mining Threats

Mining is the operation where devices in the blockchain system authorize and approve transactions. Mining threats include cryptojacking, selfish mining, security vulnerabilities, and resource depletion. In resource depletion, malicious crypto miners can

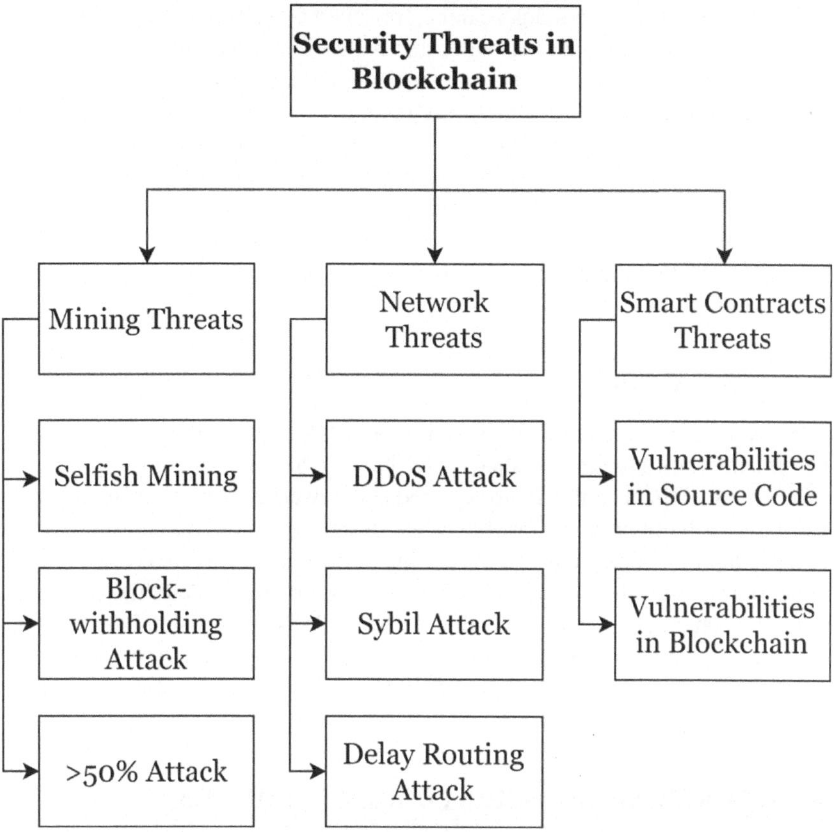

FIGURE 5.1 Security threats in blockchain networks.

utilize a system's computational power without the consumer's knowledge, which leads to system slowdowns and high power consumption. Miners get recently minted tokens in return for their computational attempt. Crypto mining malware has been identified as one that can bypass antivirus programs. There is no security mechanism against illegitimate crypto mining. The links may be in the spam mail, infecting websites with crypto-mining droppers. A managed spam filter service can differentiate spam emails from normal ones.

5.3.2 Network Attacks

DDoS Attack: A DDoS attack disturbs the data transmission to the end system, and the attackers prevent the end devices from being unavailable to authorized users. A DDoS attack is overly complex to identify due to the difficulties in distinguishing between authorized and malicious requests.

Sybil Attack: A Sybil attack is a vicious crime performed to control a blockchain network by generating several false identities to trick network users. The objective is to gain more power in the network and undermine its authority. The Sybil attacks can be prevented by selecting the accurate and most appropriate consensus algorithm for the blockchain network and monitoring other nodes that send blocks from individual users.

Routing Attack: Routing attacks are related to blockchain network security where malicious activities occur, and the objective is to interrupt the routing of data packets across the internet. An intruder can separate a network into two or more distinct components using routing attacks. Routing attacks can be rectified by implementing powerful encryption before transmitting data, using secure routing protocols, and creating awareness about threats.

5.3.3 Smart Contract Threats

Smart contracts are system codes saved on a blockchain network and can be compiled when conditions are satisfied. Intruders try to insert malicious code into a smart contract, which results in unexpected and unwanted behaviour in the contract. Security must be the highest priority when designing and developing a smart contract. The harshness of security vulnerabilities to smart contracts is apparent in the number of losses in smart contract hacks. Smart contract vulnerabilities pose a major risk to blockchain ecosystems and digital assets. These attacks can be prevented by implementing secure software development practices throughout the contract's lifecycle and setting up monitoring tools to detect the contracting activity and receive alerts for any abnormal behaviour.

5.4 SECURITY MECHANISMS OF BLOCKCHAIN TECHNOLOGY

The major concern in blockchain networks is sharing data and transactions. The various security requirements of blockchain networks are authentication, access control, and confidentiality.

1. **Authentication:** Authentication in blockchain is an essential security mechanism for identifying and verifying users online. The hashing process ensures that the data has not been tampered with, verifying the authenticity of the transaction. Cryptographic keys and digital signatures are utilized in blockchain to authenticate users, devices, and systems and prevent identity theft and fraud. Routing protocols authenticate the route, allowing some data before another router receives its routing updates.
2. **Access Control:** Access control is a mechanism for distributing rights to the parties to gather data and resources. Access control is related to choosing the permitted actions of authorized participants and arbitrating each effort by a user to access the data in the device. The objective of access control is to prevent the system resources from unauthorized user access.
3. **Digital Signature:** The digital signature validates blockchain network transactions, specifically to authenticate transactions. When users perform

transactions, they must confirm to every node in the system that they are authorized to pay those amounts while preventing other users from spending them.

4. **Key Management:** Key management is an essential part of cryptography in dealing with security services, including confidentiality of transactions, authentication, and data integrity. The private keys of authorized parties can be securely stored to avoid leakage and stealing in blockchain networks. Two conventional techniques to realize public key infrastructure are centralized by a certificate authority and decentralized by a web of trust.
5. **Data Privacy:** Data privacy in the blockchain is a major concern, and all parties on the network may access any personal or sensitive messages recorded. Data privacy concerns should be addressed to guarantee and protect the user's confidential information from unauthorized access to blockchain networks. Smart contracts can be susceptible to security threats and misuse, which results in the leakage of sensitive information stored on the blockchain. The distributed characteristics of blockchain act against data breaches and other security vulnerabilities. Zero-knowledge proof is an immense approach to defend data privacy in a blockchain system. In this technique, a person can confirm to others that a given declaration is true without revealing any further information. Different consensus mechanisms are used in blockchain technology. Proof of Work (PoW) eliminates the need for a central authority like a bank, business, or government to monitor and supervise transactions and their corresponding accounts. The PoW algorithm verifies thousands of transactions to ensure the total history of transactions remains perfect and not modified.

   Proof of activity guarantees that all transactions on the blockchain are legitimate and that all miners arrive at a consensus. When the block is mined, authorities with stakes validate and sign the block. Proof of authority is a consensus scheme that allows fast transactions using identity and money as stakes at the cost of decentralization. Proof of Authority influences the value of identities in a private network. The proof of burn algorithm can provide network security as it becomes severe for attackers. When the participants burn more coins, they have more opportunities to mine the next block, which increases the number of awards. Proof of Elapsed Time is a consensus algorithm in a permissioned blockchain network to choose mining rights and the next block miner. The proof of elapsed time algorithm selects who receives to mine coins and who wins a block using an arbitrarily generated amount of time.

5.5 DECENTRALIZATION

Decentralization in blockchain distributes the rights, control, and decision-making power among several users within a blockchain network. It guarantees that an individual, group, or government does not manage the blockchain. Blockchain provides an efficient way to serve unidentified parties without trusting each other.

5.6 CONSENSUS MECHANISMS

Consensus mechanisms provide a set of rules and protocols that manage how agreement is reached and how transactions are legalized within the network. The consensus mechanism keeps the blockchain's security through data maintenance.

5.7 NODE SECURITY

A blockchain node is a system that together compiles the blockchain's software. It facilitates the blockchain in legalizing transactions and maintaining the network's security by ensuring it remains decentralized. In order to avoid these attacks, it is essential to safeguard the node by creating strong security strategies, such as encryption, firewalls, and secure hosting environments.

5.8 SMART CONTRACTS

It serves similarly to the conventional contract while also automatically enforcing the contract. Smart contracts are programs that compute precisely as their designers program them. Smart contracts can also be used to accomplish complicated corporate practices. A smart contract may automate the tracking of products from the producer to the customer in supply chain management. The smart contract revises the ledger each time the products are exchanged, assuring that everyone in the supply chain has the right to identical information. Digital identity authentication may also be achieved with smart contracts.

5.9 ENDPOINT SECURITY

The participants of blockchain networks can directly access the network's endpoint on electronic devices such as computers and mobile phones. Intruders target the end systems, hack critical information, and make the resources unavailable to authorized parties. The endpoint security solution examines and secures each operational endpoint in a network. This is attained through a centralized management console installed on the corporate network or server. These endpoint security tools provide features such as susceptible endpoint detection, multi-factor authentication, real-time monitoring, user behavioural analysis, and others to identify advanced security threats.

5.10 CRYPTOGRAPHIC KEYS

Cryptography is a technique for protecting data against unauthorized access. The transactions between two nodes in the blockchain are secured using cryptography. The data transmitted across the peer-to-peer networks is encrypted, and hashing is utilized to secure block messages and connect blocks in the blockchain. Blockchain utilizes asymmetric cryptography to secure all transactions stored in the block. Cryptographic keys are essential for operations that take place in blockchain networks. It is necessary to create, utilize, and store the secure key in the blockchain technology. Public-key cryptography is a security mechanism that provides digital authentication to the participants. Any user can decrypt coded information transmitted to them and create a digital signature from it for authentication. A public key is open for the network, which others use to validate a transaction and encrypt a message. Authentication is another security mechanism for system login, bank account transactions, and blockchain network security. Authentication protocols make use of cryptography to validate the participant's identification. The storage and evaluation of electronic signatures is accomplished through cryptography.

5.11 KEY ESTABLISHMENT FOR BITCOIN

Bitcoin is identical to the account where the currency is maintained. The owner should have a private key to utilize the currency. Intruders can hack into the Bitcoin currency through physical or side-channel attacks. An efficient key management scheme is necessary to safeguard the Bitcoin wallet.

Password-Based Wallet: The private keys can be preserved by users who create a password. Bitcoin software only provides access to the stored keys when the right password is supplied. Intruders try to steal private keys through brute force attacks. When users forget the password, they cannot access their Bitcoin wallet.

Hosted Wallet: Users can access the services of the Bitcoin wallet through a web authentication mechanism. The third party controls the wallet security, and it is suggested to the user that a smaller amount should be maintained in the wallet.

5.12 FUTURE RESEARCH

The drawback of blockchain technology is that transactions are overly expensive since most blockchain networks utilize PoW and proof of stake. Federated learning should enhance data privacy and communication security of local devices. Scalable federated learning can deal with multiple devices and handle large data sets. Federated learning may also speed up blockchain application development, enhance blockchain security and performance, and facilitate new artificial intelligence fusion applications.

5.13 CONCLUSION

This proposed work provides a detailed analysis of security challenges, security threats, and various security mechanisms of blockchain technology. The various security mechanisms of blockchain are addressed. The various consensus mechanisms are discussed in this work, and their performance parameters are compared. This work also discussed the use of cryptographic keys in securing blockchain networks and the key management of the Bitcoin wallet. Finally, the challenges and future research of blockchain technology are discussed. Blockchain enhances information security, transparency, trust, and traceability across business networks.

Blockchain security built it to become a popular technology for several applications such as data storage, supply chain management, and financial transactions. Future research on blockchain security will involve the implementation of quantum-resistant consensus algorithms. Blockchain security can be strengthened through various quantum-resistant cryptographic algorithms such as code-based cryptography, lattice-based cryptography, hash-based cryptography, and multivariate cryptography. The quantum-resistant algorithm will soon play a key role in blockchain security.

REFERENCES

1. Shishu Ding, Hao Hu, Feng Xu, Zhenyu Chai, Wen Wang, "Blockchain-Based Security-Minded Information Sharing in Precast Construction Supply Chain Management with Scalability, Efficiency and Privacy Improvements," *Automation in Construction*, Volume 168, p. 105698, 2024.
2. Hari Mohan Rai, Kaustubh Kumar Shukla, Lilia Tightiz, Sanjeevikumar Padmanaban, "Enhancing Data Security and Privacy in Energy Applications: Integrating IoT and Blockchain Technologies," *Heliyon*, Volume 10, No. 19, p. 38917, 2024.
3. Xiaohua Wu, Jing Wang, Tingbo Zhang, "Integrating Fully Homomorphic Encryption to Enhance the Security of Blockchain Applications," *Future Generation Computer Systems*, Volume 161, pp. 467–477, 2024.
4. Selman Hızal, A.F.M. Suaib Akhter, Unal Cavusoglu, Devrim Akgun, "Blockchain-Based IoT Security Solutions for IDS Research Centres," *Internet of Things*, Volume 27, pp. 101307, 2024.
5. Zhiwei Wang, Qingqing Chen, Lei Liu, "Permissioned Blockchain-Based Secure and Privacy-Preserving Data Sharing Protocol," *IEEE Internet of Things Journal*, Volume 10, Issue 12, pp. 10698–10707, 2023.
6. Debashis Das, Sourav Banerjee, Pushpita Chatterjee, Uttam Ghosh, Utpal Biswas, "A Secure Blockchain Enabled V2V Communication System Using Smart Contracts," *IEEE Transactions on Intelligent Transportation Systems*, Volume 24, Issue 4, pp. 4651–4660, 2023.
7. Abdullah Ayub Khan, Sami Bourouis, M. M. Kamruzzaman, Myriam Hadjouni, Zaffar Ahmed Shaikh, Asif Ali Laghari, Hela Elmannai, Sami Dhahbi, "Data Security in Healthcare Industrial Internet of Things with Blockchain," *IEEE Sensors Journal*, Volume 23, Issue 20, pp. 25144–25151, 2023.
8. Zhou, Youliang Tian, Jinbo Xiong, Jianfeng Ma, Changgen Peng, "Blockchain-Enabled Secure and Trusted Federated Data Sharing in IIoT," *IEEE Transactions on Industrial Informatics*, Volume 19, Issue 5, pp. 6669–6681, 2023.
9. Shalli Rani, Himanshi Babbar, Gautam Srivastava, Thippa Reddy Gadekallu, Gaurav Dhiman, "Security Framework for Internet-of-Things based Software-Defined Networks using Blockchain," *IEEE Internet of Things Journal*, Volume 10, Issue 7, pp. 6074–6081, 2023.
10. A. Sasikumar, Logesh Ravi, Ketan Kotecha, Ajith Abraham, Malathi Devarajan, Subramaniyaswamy Vairavasundaram, "A Secure Big Data Storage Framework Based on Blockchain Consensus Mechanism with Flexible Finality," *IEEE Access*, Volume 11, pp. 56712–56725, 2023.
11. Mohd. Aquib Ansari, Dushyant Kumar Singh, "Human Detection Techniques for Real Time Surveillance: A Comprehensive Survey," *Multimedia Tools and Applications*, Volume 80, pp. 8759–8808, 2021. https://doi.org/10.1007/s11042-020-10103-4.
12. Xiaohan Hao, Wei Ren, Yangyang Fei, Tianqing Zhu, Kim-Kwang Raymond Choo, "A Blockchain-Based Cross-Domain and Autonomous Access Control Scheme for Internet of Things," *IEEE Transactions on Services Computing*, Volume 16, Issue 2, pp. 773–786, 2023.

13. Rupesh Kumar Dewang, Mahendra Pratap Yadav, Surbhit Awasthi, Om Raj, Arvind Mewada, Kamlakant Laxman Bawankule, "Data Secure Application: An Application That Allows Developers to Store User Data Securely Using Blockchain and IPFS," *Multimedia Tools and Applications*, Volume 83, Issue 15, pp. 45491–45517, 2024.
14. Anusha Vangala, Ashok Kumar Das, Ankush Mitra, Sajal K. Das, Youngho Park, "Blockchain-Enabled Authenticated Key Agreement Scheme for Mobile Vehicles-Assisted Precision Agricultural IoT Networks," *IEEE Transactions on Information Forensics and Security*, Volume 18, pp. 904–919, 2022.
15. Huaqun Guo, Xingjie Yu, "A Survey on Blockchain Technology and Its security," *Blockchain: Research and Applications*, Volume 3, pp. 1–15, 2022.
16. Jie Cui, Fenqiang Ouyang, Zuobin Ying, Lu Wei, Hong Zhong, "Secure and Efficient Data Sharing Among Vehicles Based on Consortium Blockchain," *IEEE Transactions on Intelligent Transportation Systems*, Volume 23, Issue 7, pp. 8857–8867, 2022.
17. Jianhong Zhou, Gang Feng, Yunxiang Wang, "Optimal Deployment Mechanism of Blockchain in Resource-Constrained IoT Systems," *IEEE Internet of Things Journal*, Volume 9, Issue 11, pp. 8168–8177, 2022.
18. Rui Zhang, Rui Xue, Ling Liu, "Security and Privacy for Healthcare Blockchains," *IEEE Transactions on Services Computing*, Volume 15, Issue 6, pp. 3668–3686, 2022.
19. Xu Yang, Xue Chao, Xun Yi, Ibrahim Khalil, Xiaotong Zhou, Debiao He, Xinyi Huang, "Blockchain-Based Secure and Lightweight Authentication for the Internet of Things," *IEEE Internet of Things Journal*, Volume 9, Issue 5, pp. 3321–3332, 2022.
20. Song Tang, Zhiqiang Wang, Jia Dong, Yandong Ma, "Blockchain-Enabled Social Security Services using Smart Contracts," *IEEE Access*, Volume 10, pp. 73857–73870, 2022.
21. Yong Wang, Aiqing Zhang, Peiyun Zhang, Youyang Qu, Shui Yu, "Security-Aware and Privacy-Preserving Personal Health Record Sharing Using Consortium Blockchain," *IEEE Internet of Things Journal*, Volume 9, Issue 14, pp. 12014–12028, 2022.
22. Saurabh Singh, A. S. M. Sanwar Hosen, Byungun Yoon, "Blockchain Security Attacks, Challenges, and Solutions for the Future Distributed IoT Network," *IEEE Access*, Volume 9, pp. 13938–13959, 2021.
23. Bhabendu Kumar Mohanta, Debasish Jena, Somula Ramasubbareddy, Mahmoud Daneshmand, Amir H. Gandomi, "Addressing Security and Privacy Issues of IoT Using Blockchain Technology," *IEEE Internet of Things Journal*, Volume 8, Issue 2, pp. 881–888, 2021.
24. Li Da Xu, Yang Lu, Ling Li, "Embedding Blockchain Technology into IoT for Security: A Survey," *IEEE Internet of Things Journal*, Volume 8, Issue 13, pp. 10452–10473, 2021.
25. Elham A. Shammar, Ammar T. Zahary, Asma A. Al-Shargabi, "A Survey of IoT and Blockchain Integration: Security Perspective," *IEEE Access*, Volume 9, pp. 156114–156150, 2021.
26. Kai Peng, Meijun Li, Haojun Huang, Chen Wang, Shaohua Wan, Kim-Kwang Raymond Choo, "Security Challenges and Opportunities for Smart Contracts on the Internet of Things: A Survey," *IEEE Internet of Things Journal*, Volume 8, Issue 15, pp. 12004–12020, 2021.
27. Lakshmana Kumar Ramasamy, Firoz Khan, Agbotiname Lucky Imoize, JoshuaOgbebor, Seifedine Kadry, Seungmin Rho, "Blockchain-based Wireless Sensor Networks for Malicious Node Detection: A Survey," *IEEE Access*, Volume 9, pp. 128765 – 128785, 2021..
28. Nagendra Singh, Harsh Pratap Singh, Anuprita Mishra, Anula Khare, Mamta Swarnkar, Shekh Kulsum Almas, "Blockchain Cloud Computing: Comparative Study on DDoS, MITM and SQL Injection Attack," In *2024 IEEE International Conference on Big Data & Machine Learning (ICBDML)*, Bhopal, India, 2024, pp. 73–78. https://doi.org/10.1109/ICBDML60909.2024.10577412.

29. Guangxia Xu, Yong Liu, Prince Waqas Khan, "Improvement of the DPoS Consensus Mechanism in Blockchain Based on Vague Sets," *IEEE Transactions on Industrial Informatics*, Volume 16, Issue 6, pp. 4252–4259, 2020.
30. Peiyun Zhang, Mengchu Zhou, "Security and Trust in Blockchains: Architecture, Key Technologies, and Open Issues," *IEEE Transactions on Computational Social Systems*, Volume 7, Issue 3, pp. 790–801, 2020.
31. Nagendra Singh, Yogendra Kumar, "Security and Privacy Requirements in 5G Healthcare," In *Block Chain for 5G Healthcare Applications*, IET Publication, pp. 1–30, 2021. https://doi.org/10.1049/PBHE035E_ch1.
32. Jiafu Wan, Jiapeng Li, Muhammad Imran, Di Li, Fazal-e-Amin, "A Blockchain-Based Solution for Enhancing Security and Privacy in Smart Factory," *IEEE Transactions on Industrial Informatics*, Volume 15, Issue 6, pp. 3652–3660, 2019.

Section 3

Revolutionizing Finance

6 The Transformative Impact of Blockchain Technology on the Financial Sector

Shyam Sunder Agrawal, Saloni Raheja, and Amanjot Singh Syan

6.1 INTRODUCTION

Blockchain technology is a revolutionary innovation in the digital domain, a decentralised and extremely secure method for relocating and recording data. A blockchain is a decentralised system that accomplishes a continuously mounting collection of data, organised into blocks that are linked and secured through encryption [1–3]. This decentralised structure eradicates the risk of a single point of failure, as the record is replicated and preserved by all participants in the network. In addition, the cryptographic methodsthat link the blocks make altering the data almost impossible, as the network would directly detect and reject any such attempt. The secure and decentralised countryside of blockchain has significant implications across multiple industries, including finance, supply chain management, and digital identity management. In the financial sector, blockchain technology has great promise for enabling faster, more efficient, and highly transparent transactions. It can also improve supply chain traceability, transparency, and efficiency by tracking goods and commodities in real-time [4].

Furthermore, blockchain-based digital identity solutions provide individuals better regulators over their data, ensuring confidentiality and security [5]. Blockchain differs from traditional centralised systems in that it operates on a decentralised, distributed ledger where transactions are recorded transparently and securely [6]. Blockchain technology confirms and records the transactions using a public ledger without the need for any third-party oversight or control [7]. The system arranges unified blocks in chronological order, each containing a hash of the previous block, which safeguards data integrity and makes the system resistant to meddling. Therefore, the most noteworthy application of blockchain in finance is its potential to convert the primary financial market. The traditional financial market infrastructure has long-faced challenges, but presenting blockchain technology proposals is a promising solution [4][8–10]. Its decentralised nature and improved security features can

DOI: 10.1201/9781003545620-9

address problems such as incompetent data management, absence of transparency, and reliance on trusted intermediaries [11].

6.1.1 Financial Services Blockchain Technology Challenges and Benefits

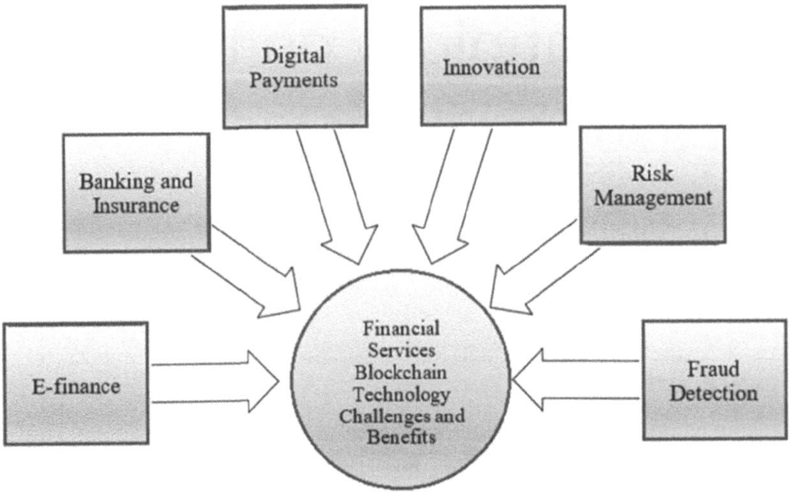

FIGURE 6.1 Financial services blockchain technology challenges and benefits.

6.1.2 Blockchain in Finance

Figure 6.1 signifies the tests and benefits of blockchain technology in various financial services, highlighting the main problems and advantages across various sectors, including e-finance, banking, insurance, digital payments, risk management, and fraud detection. Furthermore, blockchain technology has revolutionised frequent aspects of finance by enhancing transparency, privacy, and security, allowing fast settlements, modernising risk management, simplifying auditing, and reducing costs. These qualities convert traditional financial processes and organisations, creation them more efficient, secure, and accountable.

6.1.3 Emerging Trends in Blockchain Finance

Figure 6.2 illustrates the transformative impact of blockchain in finance in key areas such as transparency, privacy, security, fast settlements, risk management, auditing, and cost minimisation. These features demonstrate how blockchain technology is reshaping traditional financial processes and organisations. In addition to these progressions, blockchain has rapidly transformed various sectors, particularly finance, by presenting innovative frameworks like asset tokenisation and decentralised autonomous organisations (DAOs). This literature review synthesises recent research on these topics, emphasising their possibility to disturb and revolutionise traditional financial systems.

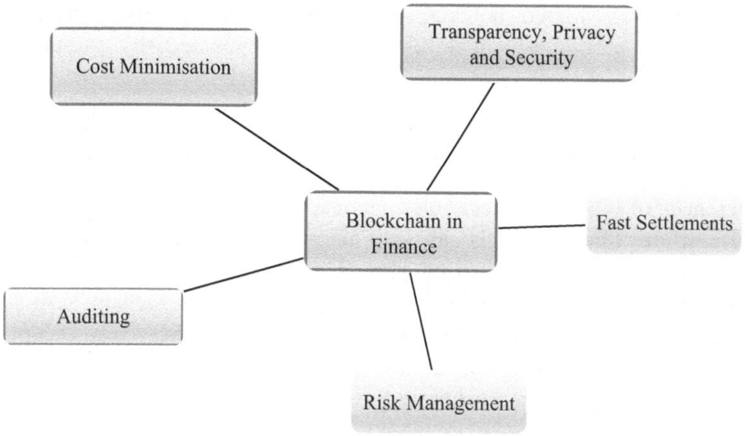

FIGURE 6.2 Blockchain in finance.

6.1.4 Tokenisation of Assets

Tokenisation is the process of changing physical or digital assets into blockchain-based tokens. This invention allows fractional ownership, recovers liquidity, and streamlines transactions. According to recent studies, tokenisation is an indispensable element in forming the World Innovation Stock Exchange, where the revolution of tokenisation and the securitisation of project management are grave (source). This suggests that tokenisation democratises entree to investment opportunities and raises better transparency and competence in asset management. Furthermore, the request for decentralised exchanges is energetic for the seamless conversation of these digital tokens and cryptocurrencies on the blockchains. The author highlighted that mixing decentralised exchanges within the decentralised finance (DeFi) ecosystems is energetic for the efficient trading of assets, thereby improving market dynamics and liquidity [12].

6.1.5 Decentralised Autonomous Organisations

DAOs signify another transformative trend in blockchain finance. These organisations operate on smart contracts and are ruled by their members, eradicating the traditional hierarchical structures. According to the researcher, DAOs have the potential to disrupt traditional funding mechanisms in various industries, particularly entertainment. Through the utilisation of blockchain tokens, DAOs can enable direct funding from stakeholders, thereby offering a more democratised method of investment and project management [13].

6.1.6 Security Challenges and Trust in Blockchain

Even though it has benefits, blockchain technology is not deprived of challenges. He recognises several security concerns, including the vulnerabilities to cyber threats, which could deteriorate the truthfulness of the financial transactions [14]. The joint of blockchain with outdated financial security instruments grants an area ready for further research, particularly in developing robust cybersecurity summaries custom-made to the blockchain environments.

6.1.7 Blockchain's Impact on Auditing and Corporate Governance

The inferences of blockchain technology spread into auditing and corporate governance. He argues that the blockchain can alter auditing through transparency and accountability in financial reporting [15]. This transformation stances questions about the future parts of the auditors and the potential need for new skill sets that contribute to blockchain knowledge. As organisations progressively accept blockchain for financial record-keeping, additional research is desirable to understand how these changes will mark the corporate governance structures and the compliance requirements.

6.2 REVIEW OF LITERATURE

Blockchain has emerged as a transformative potency in numerous industries, with finance as the lead of its adoption. Recent studies have discovered the likelihood of blockchain in finance [15]. Blockchain is a transformative technology that can improve the speed, correctness, and cost-efficiency of financial transactions, possibly leading to extensive fluctuations in how firms function and devise new ways of payment through cryptocurrencies. The financial industry has already been activated to discover and implement blockchain-based solutions, along with applications that range from cross-border payments to trade finance and asset management [6].

In addition, blockchain technology's decentralised and safe nature opens up new possibilities for financial inclusion, particularly where traditional banking services are rare [16]. It seeks to evaluate the possibility of DeFi replacing traditional finance by using the Ethereum platform enabled by blockchain. The abolition of the intermediaries and the mechanisation of the keen contracts result in cost reductions. Market share is gained by fintech companies working with blockchain startups. Limitations include industry-specific differences and long-term data availability. Implications call for more investigation into long-term effects, research particular to subsectors, and comparative evaluations. Fintech companies are urged to deliberately implement blockchain with a focus on risk management and compliance [5].

6.2.1 Transparency, Privacy, and Security

The decentralised ledger of blockchain guarantees transparency by authenticating the transactions through computers, which allows all participants to view them in real-time. Every transaction is tightly encrypted and cannot be improved in the past without the participants' agreement. This transparency nurtures trust among the parties and diminishes the risk of deceit and employment. Blockchain technology is transformative in reshaping society and industry because of the fintech revolution and web utilisation [17]. According to this research, blockchain execution has been accepted to foster transparency and trust across various sectors due to increased competition in the digital landscape [18]. Integrating blockchain and artificial intelligence has changed business processes, increased efficiency and transparency and provided security to businesses [19]. Moreover, blockchain improved financial inclusion through various solutions, particularly in developing countries, and it can be made imaginable through IoT.

6.2.2 Fast Settlements

Blockchain tolerates fast settlements, avoiding the need for intermediaries and clearinghouses commonly created in old-fashioned financial systems. Smart contracts allow transactions to be automatically verified and settled founded on predefined conditions, as the agreement terms have been inscribed into the code. This automation reduces the time and cost related to the manual verification and processing.

6.2.3 Risk Management

Blockchain recovers risk management in finance, assuming that it has transparent and indisputable records of transactions. Financial institutions can use custom blockchain-based platforms to display and scrutinise transactions in real-time by more competently identifying the possible risks and anomalies.

6.2.4 Auditing

Auditing has become more efficient and dependable with blockchain technology. Auditors can authenticate the truthfulness of financial records without trusting third parties to settle the records.

6.2.5 Cost Minimisation

Blockchain weakens the costs in finance by eliminating intermediaries and eliminating manual processes. Smart contracts systematise the transaction clearances and implement the contract terms without the essential intermediaries such as banks or clearinghouses, thereby dipping the transaction fees and operational costs.

6.3 RESEARCH METHODOLOGY

The process begins with the expression of the key research questions designed to influence the primary advantages of blockchain technology in the respective financial sector, categorising the difficulties and constraints related to its implementation, measuring its influence on the various financial services, and discovering the emerging trends. Subsequently, a complete search strategy is developed using academic resources such as IEEE Xplore, Google, Science Direct, SpringerLink, and JSTOR. Keywords such as "Blockchain," "Machine Learning," "Finance Industry," and "Regulatory" are working to answer the search. Specific inclusion criteria (IC) are recognised to select the works for presence in the review. These standards comprise keywords such as "Blockchain," "fintech," "distributed ledger technology," and

"Blockchain technology adoption" for IC 1. Rational operators are applied to define the standards' composition.

Furthermore, the studies published before December 31, 2019, are omitted (IC 2), although only the research publications in English are measured (IC 3). The study types are restricted to journal articles (IC 4), and studies based on abstracts are included (IC 5). Furthermore, studies based on full texts are incorporated (IC 6). To eliminate literature from consideration for this systematic literature review, exclusion criteria (EC) are applied. Duplicate studies with identical titles and/or digital object identifiers are removed, and studies are eliminated based on quality ratings as specified in the exclusion conditions. At this stage, 42 studies were identified, of which 27 were deemed relevant to this investigation. The selected papers considered for this study are itemised in Table 6.1.

6.4 RESULTS, DISCUSSION, AND FINDINGS

Based on the above discussion and systematic literature review, it is apparent that most research employs both qualitative and quantitative analyses. The qualitative studies are analysed using the systematic literature review method. At the same time, the quantitative analyses use various statistical software and tools such as SPSS AMOS and methods such as descriptive and inferential analysis. Numerous important discoveries are highlighted in the comprehensive literature study on the influence of blockchain technology on the respective financial sector. Blockchain technology power facilitates increased efficiency and decreased expenses by removing the mediators and restructuringthe financial operations.

6.5 IMPLICATIONS OF BLOCKCHAIN IN FINANCE

Blockchain technology can have a number of important repercussions for the financial sector. One of the most noteworthy aspects is the potential to revolutionise the payment systems. Blockchain technology permits individuals to conduct transactions directly with one another without the need for intermediaries. This results in a growth in transaction speed and a decrease in transaction costs. An additional benefit of this decentralised model is that it increases transparency and reduces the risk of fraud, which is an important concern in outdated financial systems.

Moreover, blockchain can justify the procedure of securities trading and settlement. By using keen contracts, the blockchain platforms allow the automatic execution of contracts once predefined conditions are met, eliminating the need for mediators such as clearing houses and dipping settlement times. This capability improves efficiency and drops the functioning risks and costs associated with the post-employment processes.

6.6 CONCLUSION

Blockchain technology has enhanced the financial sector by creating a secure and decentralised platform for various financial claims. There is an opportunity for the introduction of FinTech explanations based on blockchain technology to revolutionise

TABLE 6.1
SLR of the Studies

Authors	Year	Objectives	Key Findings	Challenges Identified
Chang et al. [19]	2020	To assess blockchain's impact on financial market efficiency	Blockchain improves market transparency and reduces transaction costs, but the technology is still in the early adoption phase.	Market volatility, regulatory challenges
Cucari et al. [20]	2021	Blockchain in traditional banking (Interbank Spunta project)	Blockchain improves data transparency, faster execution, and direct transfer of checks and money	Scalability, data availability, and regulatory clarity are needed.
Dong et al. [14]	2021	To analyse how Blockchain enhances transaction transparency	Blockchain provides an immutable and transparent ledger of transactions, reducing the potential for fraud and increasing participant trust.	Scalability issues, regulatory acceptance
Gadekallu et al. [21]	2022	Decentralised finance (DeFi) via Ethereum	DeFi provides tokenisation, cross-border payments, cost reductions, and improved efficiency.	Scalability, interoperability, and regulatory unpredictability.
Ghosh [22]	2020	To investigate blockchain's role in Anti-Money Laundering (AML) efforts	Blockchain enhances AML efforts by providing traceable transaction records and improving compliance with regulatory requirements.	Privacy issues, potential misuse for illegal activities
Maidamwar and Chavhan [7]	2020	To examine the rise of decentralised finance (DeFi)	DeFi platforms offer innovative financial products but face significant risks related to smart contract vulnerabilities and regulatory scrutiny.	Smart contract risks, regulatory environment
Mishra and Kaushik [11]	2023	To assess the potential of blockchain for CBDCs	Blockchain enables the creation of secure and efficient Central Bank Digital Currencies (CBDCs), offering benefits such as faster transactions and improved monetary policy implementation.	Legal and regulatory challenges, technological limitations
Moşteanu and Faccia [6]	2021	To assess the impact of blockchain on overall financial stability	Blockchain can enhance financial stability by reducing systemic risks, increasing transparency, and improving the resilience of financial systems.	Regulatory challenges, potential for new forms of systemic risks
Agrawal et al. [16]	2024	To explore the impact of artificial intelligence on the evolution of fintech	Artificial intelligence has enabled fintech startups to develop innovative solutions	Challenges to the traditional banking model

the distribution of financial services, which will result in improved accessibility, reduced expenses, and greater scrutiny. Various blockchain applications have emerged across numerous industries, including healthcare, the Internet of Things, security, and supply chain management. This is because blockchain has made it possible to construct a storage system that is decentralised, secure, and conspicuous. Blockchain technology has the potential to transform many different banking procedures, which means that it might significantly impact the financial sector. The power of blockchain technology has revolutionised the financial industry, giving rise to DeFi. This new method of financial services is shadowed by its decentralised countryside, innovative solutions, seamless interoperability, global reach, and supreme transparency. By approving DeFi, we can temporary financial inclusion, encourage open access, stimulate unrestricted novelty, and unlock the world of options for entrepreneurs and visionaries. From the industry perspective, blockchain technology significantly enhances efficiency by streamlining transactions, improving security through decentralisation, and fostering transparency with a tamper-proof ledger.

REFERENCES

1. Abdelmaboud, A., Ahmed, A. I. A., Abaker, M., Eisa, T. A. E., Albasheer, H., Ghorashi, S. A., & Karim, F. K. "Blockchain for IoT applications: taxonomy, platforms, recent advances, challenges and future research directions", *Electronics*, 11(4), 630, 2022.
2. Singh, N., & Kumar, Y. "Security and privacy requirements in 5G healthcare", In *Block Chain for 5G Healthcare Applications*, IET Publication, pp. 1–30, 2021. https://doi.org/10.1049/PBHE035E_ch1.
3. Singh, N., Sasirekha, S. P., Dhakne, A., Sai Thrinath, B. V., Ramya, D., & Thiagarajan, R. "IOT enabled hybrid model with learning ability for E-health care systems", *Measurement: Sensors*, 24, 1–14, 2022. https://doi.org/10.1016/j.measen.2022.100567.
4. Lal, R., Chhabra, A., Singla, S., & Sharma, D. "Blockchain technology: revolutionizing trust, transparency, and transaction efficiency", In *2024 International Conference on Knowledge Engineering and Communication Systems (ICKECS)*, Chikkaballapur, India, pp. 1–5, 2024.
5. Singh, N., Singh, H. P., Mishra, A., Khare, A., Swarnkar, M., & Almas, S. K. "Blockchain cloud computing: comparative study on DDoS, MITM and SQL injection attack", In *2024 IEEE International Conference on Big Data & Machine Learning (ICBDML)*, Bhopal, India, 2024, pp. 73–78, https://doi.org/10.1109/ICBDML60909.2024.10577412.
6. N. Moşteanu, & A. Faccia, Fintech frontiers in quantum computing, fractals, and blockchain distributed ledger: paradigm shifts and open innovation. *Journal of Open Innovation: Technology, Market, and Complexity,* 2021.
7. P. Maidamwar, & N. Chavhan, "Blockchain technology: a review on architecture, security issues and challenges. *IJEAST,* 04(12), 245–249, 2020.
8. da Cunha, P. R., Soja, P., & Themistocleous, M. "Blockchain for development: a guiding framework", *Information Technology for Development*, 27(3), 417–438, 2021.
9. Dinesha, D., & Balachandra, P. "Decentralised token exchanges in blockchain-enabled interconnected smart microgrids", In *2023 IEEE Belgrade PowerTech*, Belgrade, Serbia, pp. 1–6, 2023.
10. Dewang, R. K., Yadav, M. P., Awasthi, S., Raj, O.. Mewada, A., & Bawankule, K. L. "Data secure application: an application that allows developers to store user data securely using blockchain and IPFS", *Multimedia Tools and Applications*, 83(15), 45491–45517, 2024.

11. L. Mishra, & V. Kaushik, "Blockchain is used to deal with sustainability issues and challenges in the financial sector". *Journal of Sustainable Finance and Investment,* 13(3), 1318–1333, 2023.
12. Karthikeyan, D., Amutha, B., Devi, K. M., Singh, N., & Anglin Kuruba, P. "Hybrid rater to quantify and measure the severity of infection and spread of infection in musk-melon", *Bulletin of Electrical Engineering and Informatics (BEEI)*, 133, 1667~1675, 2024. https://doi.org/10.11591/eei.v13i3.5432.
13. Ansari, M. A., & Singh, D. K. "Human detection techniques for real time surveillance: a comprehensive survey", *Multimedia Tools and Applications*, 80, 8759–8808, 2021. https://doi.org/10.1007/s11042-020-10103-4.
14. L. Dong, Y. Qiu, F. & Xu. "Blockchain-enabled deep-tier supply chain finance". *Manufacturing & Service Operations Management,* 25(6), 2021-2037, 2023.
15. Mishra, L., & Kaushik, V. "Blockchain is used to deal with sustainability issues and challenges in the financial sector", *Journal of Sustainable Finance and Investment,* 13(3), 1318–1333, 2023.
16. Agrawal, S. S., Rose, N., Sahai, K. P., Maheshwari, A., & Josyula, H. P. "The FinTech revolution: AI's role in disrupting traditional banking and financial services", *Decision Making: Applications in Management and Engineering*, 10(1), 15–29, 2024.
17. Alzoubi, M. M. "Investigating the synergy of Blockchain and AI: enhancing security, efficiency, and transparency", *Journal of Cyber Security Technology*, 10, 1–29, 2024.
18. Bennet, D., Maria, L., Putri Ayu Sanjaya, Y., & Rahmania Az Zahra, A. "Blockchain technology: revolutionising transactions in the digital age", *ADI Journal on Recent Innovation*, 5(2), 192–199, 2024. https://doi.org/10.34306/ajri.v5i2.1065.
19. Chang, S. E., Luo, H. L., & Chen, Y. "Blockchain-enabled trade finance innovation: a potential change in basic assumptions on using the letter of credit", *Sustainability*, 12(1), 1–16, 2020.
20. Cucari, N., Lagasio, V., Lia, G., & Torriero, C. "The impact of blockchain in banking processes: the Interbank Spunta case study", *Technology Analysis & Strategic Management*, 34(2), 138–150, 2021.
21. Gadekallu, T., Huynh-The, T., Wang, W., Yenduri, G., Ranaweera, P. S., Pham, V. Q., da Costa, D. B., & Liyanage, M. "Blockchain for the metaverse: a review", ArXiv, abs/2203.09738, 2022.
22. K. Ghosh. "Blockchain Technology: A Catalyst for Change in Finance". *Drishtikon: A Management Journal,* 12(1), 2020.

7 Significance of Cryptography Techniques in Blockchain

Satyam Omar, Namita Tiwari, Vinay Shukla, Manvi Prajapati, and Swati Rawal

7.1 INTRODUCTION

Blockchain is a distributed ledger technology that enables decentralised data storage across several computers (nodes). Blockchain comprises a block's series, and each block plays the role of a public ledger, having a list of transactions, a timestamp, and a link to the block before it. Because of this structure, it is very difficult to change the data without also changing all of the blocks that come after it, which would require agreement from most of the network. Since a network of nodes uses a consensus method to validate and record transactions, the decentralised nature of blockchain eliminates the need for central authority. This decentralisation guarantees the security, integrity, and transparency of the data recorded on the blockchain when combined with cryptographic methods [1].

For decades, the study and application of secure communication methods, or cryptography, has been a vital component of human civilisation. To be more precise, cryptography is a means of securely communicating via an unreliable channel [2]. Today, cryptography serves as the foundation for many different digital security systems, guaranteeing data integrity, confidentiality, and authenticity. Blockchain, which mostly depends on encryption, is one of the most important inventions of the 21st century. Cryptography has a historical background. For instance, the Egyptians used substitution ciphers to secure important data on papyrus scrolls. After that, the Greeks made a noticeable contribution to the initial development of cryptography [3].

Later, Julius Caesar was known for deploying Caesar's cipher to encode conversations in the military. All these strategies were heavily based on complex decipherability and secrecy. In the Middle Ages, the creation of poly-alphabetic cypher advanced cryptography significantly. The 19th century was important from the point of view of cryptanalysis, a science of breaking codes [4]. Charles Babbage's work on automated decryption apparatus and Friedrich Kasiski's frequency analysis set the background for modern cryptanalysis techniques. In the modern age, symmetric ciphers such as **Data Encryption Standard** (DES) and **Advanced Encryption Standard** (AES) were designed by deploying fiestel and substitution-permutation networks, respectively. In 1976, Whitfield Diffie and Martin Hellman [1] developed

DOI: 10.1201/9781003545620-10

a key exchange protocol that opened the gateway of public-key cryptography. Later, the **Rivest–Shamir–Adleman** (RSA) cryptosystem [5] was designed, which comprises two distinct keys, namely the public key for encryption and the private key for decryption. After that, the fields of information security were completely transformed by ElGamal [3], **Digital Signature Algorithm** (DSA), and **Elliptic Curve Digital Signature Algorithm** (ECDSA). These innovations revolutionised applications like digital signatures, secure email, and safe online transactions by enabling secure communication without the requirement for a shared secret key.

Blockchain, a decentralised and distributed ledger technology, contributes to various sectors by providing secure, transparent, and immutable transaction records. Cryptographic protocols are important tools in accomplishing this target for blockchains. This chapter investigates the significance of cryptography in securing blockchain, looking at the various methods and applications through which the blockchain's integrity and security could be customised by cryptographic techniques [6].

7.2 ROLE OF CRYPTOGRAPHY IN BLOCKCHAIN SECURITY

Several critical components of blockchain security depend on cryptography, such as data confidentiality, data integrity, data authenticity, non-repudiation, consensus mechanisms, digital signatures, hash functions, and zero-knowledge proofs (ZKPs).

7.2.1 Data Confidentiality

Data confidentiality secures the data from unauthorised access [7]. To maintain confidentiality, blockchain-primarily depends on encryption techniques. Encryption is a cryptographic technique that changes the original message into an unintelligible text that can only be retrieved in the original format by a special keyholder. Depending on the number of keys, the encryption techniques can be classified into the following two categories.

1. **Symmetric Encryption:** In this type of technique, only one key is used by the sender of the message and the recipient of the message. This means that the same key is used for both encryption and decryption. As we know, $n(n-1)/2$ keys will be required to be shared in a network of n members. Therefore, despite being very effective and quick, it is not frequently employed in public blockchains. However, it is occasionally used in private blockchains to maintain confidentiality.
2. **Asymmetric Encryption:** In this type of encryption technique, two distinct keys are used by the sender and receiver of the message. The sender uses the recipient's public key to encode the original message, and the recipient uses his private key to decrypt the encoded message. There are only $2n$ keys required in a network of n users. This is why asymmetric encryption is frequently employed in blockchain systems to protect participant communications and transactions. Sensitive information is kept secret because only the owner of the private key may decrypt the data.

7.2.2 Data Integrity

Data integrity ensures that the data has not been changed in an unauthorised manner [8]. The cryptographic hash function (CHF) is the most essential tool for maintaining integrity in blockchain applications. Modern digital security tools such as signature schemes, password protection, data integrity checks, and blockchain technology deploy CHFs. A CHF takes the input of flexible length and converts it into a string of smaller fixed length. The following are the key features of CHFs that make them practically usable.

1. **Deterministic Output:** A CHF always provides the deterministic output. It means that if we give the same inputs to a CHF, it will provide the same output again and again. Due to this property, CHFs are useful in data integrity verification.
2. **Fixed-Length Output:** A CHF consistently generates an output (hash) of the same length regardless of the size of the input data set. For example, whether the input is a single letter or an entire paragraph, the SHA-256 hash function always returns a fixed bit, i.e., a 256-bit (32-byte) hash result. Comparison and hash value management are made simpler by this set length.
3. **Efficiency:** Fast and effective in processing vast volumes of data to generate hashes, CHF is built to last. This performance is essential for applications such as blockchain technology and internet communication security, wherein massive volumes of data must be hashed in real-time.
4. **Pre-Image Resistance:** This property is all about finding the pre-image of a given hash output. It should be computationally impossible to deduce the original input given only the hash result. For sensitive data (for example, passwords), this characteristic is crucial when only the hash is kept.
5. **Collision Resistance:** When two distinct inputs provide the same hash output, this is known as a collision. The goal of CHFs is to reduce the possibility of collisions so that every unique input produces a unique hash. For blockchain technology and digital signatures to preserve data confidentiality and integrity, collision resistance is essential.
6. **Avalanche Effect:** This characteristic of CHFs indicates that a slight modification in the input of CHF results in a significantly different hash output. Due to this feature, it is simple to identify unauthorised changes, as even small changes to the input data result in an entirely different hash.

CHFs are extensively employed in numerous digital security sectors. To improve security in the case of a data breach, they are employed in password storage to store hashed passwords rather than the actual passwords safely. Hashing functions make it possible to quickly compare data in data integrity verification to make sure it has not been altered. Because malware infections can be disinfected by checking file integrity, it is significant for software distribution [9].

For practical usability, we can employ iterated hash functions. The iterated hash function is an example of a CHF, which takes an input message of any length and converts it into fixed-size blocks known as the hash value or digest. It repeatedly applies a compression function to each block. An iterated hash function works as follows.

1. **Message Padding:** Message padding ensures that the length of the input message exceeds the block size. The standard steps in padding are to add one 1 bit, enough 0 bits, and then the length of the original message. The objective of this phase is to ensure that the message can be appropriately divided into fixed-size blocks.
2. **Initialisation Vector:** An initialisation vector (IV), which is a fixed, specified value, is used to begin the hash computation. It is essential to the security of the hash function because it ensures that two identical messages shall produce different hash values if different IVs are utilised.
3. **Processing Blocks:** The padded message divided into fixed-size blocks is processed one block at a time. The compression function accepts the current block as input and outputs the hash of the block (or, in the case of the first block, the IV). The compression function combines these inputs to create a fixed-size output, which is then used as the input for processing the following block.
4. **Finalisation:** The hash value of the entire message is the compression function's final output once all blocks have been processed. The cryptographic digest that accurately captures the original message is this fixed-length output.

The following are well-known CHFs based on the iterated hash function model.

1. **MD5:** It takes an input of 512-bit blocks and outputs a 128-bit hash value. It is currently regarded as insecure because collision attacks happen due to some flaws.
2. **SHA-1:** It takes 512-bit input blocks and produces a 160-bit hash value as output. SHA-1 was once widely used, but because of similar security issues, specifically the possibility of collisions, it has been deprecated.
3. **SHA-256:** SHA-256 is a member of the SHA-2 family. It takes 512-bit blocks and returns a 256-bit hash value. It is still a popular and safe hash function, particularly in applications involving blockchain technology and other cryptography.
4. **SHA-3:** The structure of SHA-3 depends on the sponge construction so that it can take the input of variable length and produce the output of variable length.

7.2.3 Data Authenticity

The process of confirming that data originates from a reliable source is known as data authenticity [10]. In blockchain, digital signatures are the main means of ensuring authenticity. Public key cryptography uses two different keys: a private key and a public key. These are used to establish digital signatures. The public key is kept in the dictionary, and the private key is kept confidential. The following procedures are involved in the creation of a digital signature.

1. **Hashing the Message:** The original data runs through a CHF to generate a fixed-length hash value (also known as a message digest) prior to signing a message. The hash function guarantees that even slight modifications to the message produce a significantly different hash.
2. **Signature Process:** The signer's private key is then used to encrypt the hash value. The outcome is a digital signature that is specific to the message and the private key that was utilised. The signature acts as evidence that the signer is the original creator of the communication because only the signer has access to the private key.
3. **Verification Process:** A digital signature can be verified by the recipient by decrypting it with the signer's public key and getting the original hash value. After that, the recipient independently hashes the message they received. The communication is regarded as genuine and unaltered, and the signature is confirmed if the independently generated hash and the decoded hash match.

These are the following key properties and benefits of digital signatures.

1. **Authenticity:** You can get the assurance through the digital signature that a communication or document comes from a certain sender. The recipient can be sure that the communication is genuine because the signature can only be produced using the signer's private key.
2. **Integrity:** With digital signatures, you can be sure that a communication or document has not had its content changed after it was signed. The verification process will fail if the original message is altered in any way, as even a little change will provide a different hash value.
3. **Non-Repudiation:** When a signer cannot retract their signature, the communication or document is considered non-repudiated. A digital signature binds the signer to the material once it is created, and in many places, this might have legal consequences. In situations such as contracts and financial transactions, this is essential.

A lot of contemporary technology and procedures depend on digital signatures. They confirm the identity of the sender and assure that the email content has not been changed in secure email communications. Digital signatures in software distribution prevent bad actors from tampering with software and, therefore, allow consumers not to download and install tampered software. Digital signatures protect and validate financial transactions by allowing only valid parties to start or approve transactions [11]. Digital signatures are used to authenticate transactions using blockchain technology, and as soon as a user initiates a transaction, it is signed with his private key. This block would then connect this transaction to the network nodes using the user's public key to validate his signature. This prevents fraudulent and undue actions because only valid transactions can stay within the blockchain.

7.2.4 Non-Repudiation

The assurance that something cannot be denied is called non-repudiation [12]. Digital signature and non-repudiation are ideas that go perfectly together with the use of blockchain. A digital signature provides proof that a sender caused and authorised a transaction because it is linked only to the sender's private key. Once such a transaction has been signed and set in the blockchain, the sender cannot deny participation in the transaction. Digital signatures were already covered in the section before this one.

7.2.5 Consensus Mechanisms

Protocols known as consensus mechanisms make sure that every node is in agreement with the legitimacy of transactions and the ledger's current state [13]. These systems depend on cryptographic approaches to ensure that consensus is obtained securely and equitably.

1. **Proof of Work:** Proof of Work (PoW) [7] is the most well-known consensus algorithm employed by Bitcoin and other cryptocurrencies. In essence, it asks players or miners to solve challenging mathematical puzzles that resemble CHFs. Miners solve the riddle in addition to receiving the payment in Bitcoin. Because of the riddles' difficulty, it is hard for one party to control the blockchain because it takes a lot of computing power to create a new block. PoW depends on hash function security to thwart attacks such as double-spending, which is the practice of spending the same coin more than once.
2. **Proof of Stake:** An alternative to PoW [8] is Proof of Stake (PoS), in which the person who creates the next block is selected according to their stake (amount of cryptocurrency) in the network. PoS employs cryptographic approaches to ensure the selection process is secure and random, lowering the risk of centralisation and 51% attacks, even though it does not rely on solving cryptographic problems.
3. **Delegated PoS:** In a variant of PoS referred to as Delegated Proof-of-Stake (DPoS) [9], participants elect a limited number of delegates who are in charge of approving transactions and maintaining the blockchain. To protect the voting process and guarantee that the chosen delegates behave in the network's best interest, DPoS employs cryptographic techniques.

7.2.6 Digital Signatures and Authentication

As previously stated, digital signatures play a critical role in guaranteeing the legitimacy of data in blockchain systems [6]. The sender's private key is used to create a unique signature during the data or transaction signing process. The signature is thus stamped onto the data in such a manner that anyone would be able to verify using the sender's public key. However, verification authenticates the sender and checks if any modification has been made to the data. Digital signatures are used for authentication in smart contracts and also for securing transactions. Smart contracts are self-executing contracts that have complex contract specifications programmed in their codes. These are more transparent and need fewer middlemen as they are executed and kept on the blockchain. Digital signatures assure both the legitimacy of its parties and the security of the execution of the smart contract.

7.2.7 Hash Functions in Blockchain

CHF is one of the leading cryptographic techniques regarding blockchain technology [6], which supports procedures towards consensus while giving unique IDs and ensuring data integrity. Merkle trees [13], sometimes called hash trees, are currently part of blockchain systems as the most effective way to sort and validate large amounts of data. A Merkle tree, sometimes referred to as a hash tree, is any form of data structure used to collate and verify humongous sizes of data. This structure, named after computer scientist Ralph Merkle, takes in data to hash and repeatedly hashes it in pairs until it produces a single root hash, called the Merkle roots. In principle, this means that the Merkle root can be used as a kind of fingerprint for the set without needing to access the overall data set, and it could be used to verify any given piece of data within it. The following layers are used in the design for a Merkle tree, in which leaves are the discrete pieces of information that require hashing. Non-leaf nodes are intermediate hashes, which stand for child node pairs, and the Root node is a component that summarises the whole collection of data. It is because Merkle trees can efficiently verify data integrity in a distributed decentralised setting without having each network member keep the complete data set, thereby gaining popularity in technology for blockchain. Blockchain is a decentralised ledger that keeps records and verifies information through distributed consensus. A blockchain's Merkle tree facilitates the safe, effective record-keeping and verification of the transactions that are contained in each block. The following are the key significances of Merkle trees in blockchain:

- **Efficient Data Verification:** Blockchain networks involve a lot of transactions, and it would take a lot of resources to verify the integrity of each transaction independently. Merkle trees enable nodes to check only a subset of the data, rather than the entire blockchain, to confirm the integrity of every transaction. This process is referred to as "Merkle proof" or "Merkle verification."
- **Space Efficiency:** Each node in a blockchain network does not have to keep track of every transaction to ensure the chain's integrity. Alternatively, nodes can retain the Merkle root and use Merkle proofs to validate transactions when necessary. As a result, participating nodes use a significantly smaller amount of storage space.
- **Faster Verification:** By only requiring a logarithmic number of hashes to be examined (i.e., the height of the tree) instead of the complete list of transactions, a Merkle tree enables speedy confirmation of whether a specific transaction is part of a block. This significantly cuts down on the amount of time required for transaction verification.
- **Data Integrity and Immutability:** Hash functions employed in Merkle trees have cryptographic qualities that are crucial to the blockchain's security.

Since every level of the tree builds off the hashed outcomes of the one before it, changing even a single transaction would produce a distinct Merkle root. The tamper-evident nature of the data guarantees that any alterations are promptly detected, hence adding to the immutability of the blockchain.

7.2.8 Zero-Knowledge Proofs

ZKPs allow one party to persuade another that a statement is true without providing any secret information. ZKPs hold an interesting position in blockchain as they can address the dual issues of privacy and security. In addition, ZKPs can also support transactions where only certain details are concealed, and the parties involved are enabled to verify the validity of the transaction even if the parties' identities, along with the amount transferred, are hidden. This is an area of interest. Zcash [12] is a privacy-focused cryptocurrency that uses a flavour of ZKPs called Zero-Knowledge Suffix Non-Interactive Arguments of Knowledge (zk-SNARKs). ZKPs have also been applied in identity management systems and have made improvements to blockchain technology's privacy and security to help prevent potential fraud and identity theft risks.

7.3 CRYPTOGRAPHIC CHALLENGES IN BLOCKCHAIN SECURITY

Blockchains have a significant amount of protection due to cryptography, but they also face many challenges. Major cryptographic issues, such as quantum computing, are advancing technologies that threaten all the encryption methods used in current blockchain technology [14]. This advanced type of computing can solve some specific mathematical problems much quicker than supercomputers by using the properties of quantum mechanics. This advanced type of computing can solve certain specific mathematical problems much quicker than supercomputers by relying on properties of quantum mechanics, including breaking well-known cryptographic algorithms such as RSA and **Elliptic Curve Cryptography** (ECC) that are currently used in blockchain platforms. Because the quantum computer could break them, researchers are, therefore, starting to take a look at **Post-Quantum Cryptography** (PQC), which is resistant to quantum attacks. The long-term security of blockchain networks will thus depend on the adoption of PQC techniques in the advancements in quantum technology. Although highly secure, totally transparent, and decentralised, blockchain technology poses considerable challenges to scalability and efficiency.

The processing and validation of large amounts of transactions form a significant challenge to the growth of blockchain networks. While maintaining high-security systems for blockchain technology, the role of cryptography may indirectly affect the scalability and efficiency of the network systems. In previous discussions, we discussed the interrelationship of blockchain scalability and cryptography, the challenges faced by this problem, and how they can be overcome with advancements in cryptographic techniques. Key management is an essential component of blockchain security. The confidentiality of private keys is essential to the security of the system in asymmetric encryption. A lost or stolen private key can compromise related assets or data, resulting in large-scale financial losses or data breaches. Secure key storage solutions, multi-signature wallets, hardware security modules, and other effective key management techniques are crucial for protecting private keys and guaranteeing the security of blockchain systems.

7.4 FUTURE TRENDS IN CRYPTOGRAPHY AND BLOCKCHAIN SECURITY

As blockchain technology develops further, new ideas and cryptographic techniques are appearing to improve security and tackle the issues confronting the sector. The following are future trends in cryptography and blockchain security. The emergence of quantum computers is driving post-quantum cryptography research. Blockchain systems will require new encryption algorithms that can withstand quantum attacks in the future. The following post-quantum methods [10] can be applied in place of traditional cryptographic primitives: lattice-based cryptography, where the security depends on lattice problems such as the closest vector problem and the shortest vector problem; hash-based cryptography, where the security depends on the pre-image resistance of the hash function; code-based cryptography, where the security depends on coding theory problems such as syndrome decoding; multivariate cryptography, where the security depends on algebraic geometrical problems such as multivariate quadratic problems and isomorphism of polynomial problems; and isogeny-based cryptography, whose security relies on finding the isogenies within elliptic curves. Such algorithm development and standardisation will constitute one of the hottest topics in the coming years for both cryptography and blockchain communities.

In homomorphic encryption [11], calculations can be performed on encrypted data without first decrypting it. In this way, private, secure data processing within blockchain systems becomes possible. This approach is particularly useful for SMPC and privacy-preserving smart contracts. Enhanced privacy-conserving methodologies that go beyond ZKPs, such as differential privacy and secure multiparty computation (SMPC), are also being investigated for application within blockchain frameworks. These methodologies improve security and confidentiality for transactions, data exchange, and decentralised applications. The ever-increasing demand for privacy in the blockchain environment makes these approaches increasingly appealing.

Interoperability and cross-chain security are critical goals of cryptography, including enabling the seamless and secure transfer of both assets and information across diverse blockchains. This provides greater cross-chain communication and transaction possibilities. However, vulnerabilities in one blockchain can potentially affect others, making cross-chain security a pressing concern. Advancements in cryptographic protocols, including cross-chain bridges, atomic swaps, and others, are expected to lead to further improvements in both the security and interoperability of blockchain networks.

7.5 CONCLUSION

Blockchain technology uses the foundation of cryptography in terms of security. Cryptographic algorithms secure and decentralise the blockchain systems, which ensures data integrity, secrecy, authenticity, and non-repudiation. However, as the blockchain industry expands and advances, new risks and challenges like scalability issues and quantum computing are emerging. To solve these problems, cryptography research and innovation must continue. Post-quantum cryptography, homomorphic encryption, and advanced privacy-preserving techniques will be necessary to ensure the long-term security and viability of blockchain technology.

To sum up, cryptography is essential to blockchain security. As blockchain technology continues to transform industries and the digital world, cryptographic techniques will remain at the forefront of protecting the future of decentralised systems.

REFERENCES

1. N. Goenkaa, A. K. Sharmab, S. Tiwaric, N. Singhd, V. Yadave, S. Prabhuf and K. Chadaga, "A regularized volumetric ConvNet based Alzheimer detection using T1-weighted MRI images", *Cogent Engineering*, Vol. 11, No. 1, p. 2314872, 2024. https://doi.org/10.1080/23311916.2024.2314872.
2. N. Singh, H. P. Singh, A. Mishra, A. Khare, M. Swarnkar, and S. K. Almas, "Blockchain cloud computing: comparative study on DDoS, MITM and SQL injection attack", *2024 IEEE International Conference on Big Data & Machine Learning (ICBDML)*, Bhopal, India, 2024, pp. 73–78. https://doi.org/10.1109/ICBDML60909.2024.10577412.
3. D. Karthikeyan, B. Amutha, K. Mekata Devi, N. Singh, and P. Anglin Kuruba, "Hybrid rater to quantify and measure the severity of infection and spread of infection in muskmelon", *Bulletin of Electrical Engineering and Informatics (BEEI)*, Vol. 13, No. 3, pp. 1667–1675, 2024. https://doi.org/10.11591/eei.v13i3.5432.
4. N. Singh, S. P. Sasirekha, A. Dhakne, B. V. Sai Thrinath, D. Ramya, and R. Thiagarajan, "IOT enabled hybrid model with learning ability for E-health care systems", *Measurement: Sensors*, Vol. 24, pp.1–14, 2022. https://doi.org/10.1016/j.measen.2022.100567.
5. R. K. Dewang, M. P. Yadav, S. Awasthi, O. Raj, A. Mewada, and K. L. Bawankule, "Data secure application: an application that allows developers to store user data securely using blockchain and IPFS", *Multimedia Tools and Applications*, Vol. 83, No. 15, pp. 45491–45517, 2024.
6. P. W. Shor, "Polynomial-time algorithms for prime factorization and discrete logarithms on a quantum computer", *SIAM Review*, Vol. 41, No. 2, pp. 303–332, 1999.
7. B. S. Sriman, G. Kumar, and P. Shamili, "Blockchain technology: consensus protocol proof of work and proof of stake", In *Intelligent Computing and Applications: Proceedings of ICICA 2019*, Springer Singapore, pp. 395–406, 2021.
8. F. Saleh, "Blockchain without waste: proof-of-stake", *The Review of Financial Studies*, Vol. 34, No. 3, pp. 1156–1190, 2021.
9. S. M. S. Saad, and R. Z. R. M. Radzi, "Comparative review of the blockchain consensus algorithm between proof of stake (pos) and delegated proof of stake (DPOS)", *International Journal of Innovative Computing*, Vol. 10, No. 2, pp. 27–32, 2020.
10. D. J. Bernstein, and T. Lange, "Post-quantum cryptography", *Nature*, Vol. 549, No. 7671, pp. 188–194, 2017.
11. X. Yi, R. Paulet, E. Bertino, and X. Yi, *Homomorphic Encryption*, Melbourne: Springer International Publishing, pp. 27–46, 2014.
12. D. Hopwood, "Zcash protocol specification", *GitHub*, Vol. 32, No. 220, P. 1–224, 2016.
13. S. Jing, X. Zheng, and Z. Chen, "Review and investigation of Merkle tree's technical principles and related application fields", In *2021 International Conference on Artificial Intelligence, Big Data and Algorithms (CAIBDA)*, Xi'an, China, IEEE, pp. 86–90, 2021.
14. N. Singh, and Y. Kumar, "Security and privacy requirements in 5G healthcare", In *Block Chain for 5G Healthcare Applications*, IET Publication, pp. 1–30, 2021. https://doi.org/10.1049/PBHE035E_ch1.

8 Leveraging the Internet of Things and Blockchain for Developing Intelligent Learning Models

Shahnawaz Ahmad, Mohammed Wasid, Shahadat Hussain, Sana Shahab, and Jiyaul Mustafa

8.1 INTRODUCTION

In recent years, Internet of Things (IoT) and blockchain technology have altered the digital world by presenting new solutions across many industries. The IoT defines the integration of a contemporary device over the World Wide Web as interactant, aggregative, and cooperative via Internet-inculcated sensors and actors [1,2]. This revolution presents a smart environment where home appliances, machines, and all sorts of objects can communicate and self-regulate their activities, enhancing automation's performance, health, and scope in almost all aspects, including health, farming, production, and many others [3].

Blockchain, an open and distributed database, and IoT are the two technologies being embraced globally by organisations due to their profound capabilities in addressing some critical issues affecting IoT, such as security, privacy, and trust [4]. This problem becomes more challenging because more authenticated users join the IoT networks, hence there is an increased need to check the authenticity and integrity of data being exchanged by the devices in the system. Blockchain presents a strong base that reduces these factors' effects by enhancing secure data management, preventing the alteration of records, and using decentralised consensus techniques [5]. This makes it appropriate for use in main applications such as smart contracts, supply chains, and other decentralised finance applications because they require high-security measures as well as more visibility.

In parallel with such development, intelligent learning models (ILMs) serve as helpful tools for dealing with the huge amount of data generated by IoT devices. Artificial intelligence (AI), machine learning (ML), and deep learning (DL) build models so that the systems can learn the patterns and further make correctly targeted estimations and decisions. AI is the general concept of the capacity of a computer to perform work intelligently [6]. AutoAI, a subcategory of AI, is an executive-aiding application of algorithms that adapt in response to specific data inputs. DL is a subcategory of ML and it employs artificial neural networks with added layers to mine

 DOI: 10.1201/9781003545620-11

complex data and provide improved accuracy [7]. These four systems, which include IoT, blockchain, and ILMs, work harmoniously to enable the collection, sharing, and analysis of big data. This chapter explores how these technologies can be leveraged to develop ILMs that enhance decision-making and operational efficiency across various sectors, thereby driving innovation and solving contemporary challenges.

8.2 BACKGROUND

8.2.1 Internet of Things

The IoT refers to the interconnection via the Internet of computing devices embedded in everyday objects that enable them to send and receive data. This is possible because almost all "things" or objects can now be equipped with various sensors and technologies for collecting and exchanging data. Sensors can obtain environmental data by collecting information on light, heat, sound, chemical composition, or other similar factors. Embedded systems use this sensory information to interact with their external environment by taking appropriate actions. IoT technology allows multiple devices to communicate using web services and embedded software. Web services are intermediaries between different processes or applications within a single system or between different systems on the same platform or across multiple platforms [1].

The IoT has many applications, such as tracking animal behaviours, identifying soil moisture levels in farms, monitoring diabetes patients, improving personal healthcare management, developing greater self-assistance tools for older people, and enhancing environmental monitoring functions. It was estimated that there would be nearly 20 billion IoT devices worldwide by 2020, and global spending on IoT technology is projected to reach somewhere near $1.3 trillion. The blockchain market is projected to reach $3 trillion by 2024. Many companies, including Google, Amazon, Cisco Systems, Intel, General Electric, IBM, Dell, Microsoft, HP, and Bosch, have adopted IoT technology to enhance their operations. For example, in 2015, General Electric established GE Digital to transform industrial machines into smart machines, launching Predix.io as an IoT platform that provides machine monitoring tools for industries such as hospitality, aviation, and utilities.

Typically, IoT architecture comprises five primary layers, as shown in Figure 8.1. The first layer, perception layer, is responsible for sensing the information (data/actuators) from the surrounding physical world. It uses different technologies, such as Wi-Fi, Bluetooth, and 5G, to connect and communicate with the device machines/objects. The second layer, known as the network layer, transfers all information the perception layer collects to other layers for further processing. The third layer,

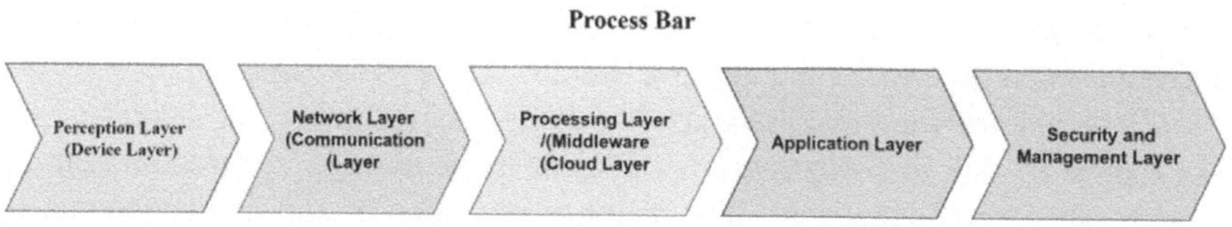

FIGURE 8.1 Layers of the IoT architecture.

processing layer, contains cloud components or middleware, which store, filter, and analyse all transferred data from the network layer. The fourth layer, application layer, provides actual data (information), i.e., it provides services to the user by creating Application Programming Interfaces (APIs) using data received from the processing layer. The fifth layer, security and management layer, handles the security and management of the entire architecture. It includes various activities such as securing data during transmission over several networks or using a time security policy so no one can use that device within the period mentioned.

A device is considered "smart" when it has two essential capabilities. First, it must be able to interact with the physical world, which involves processing collected data and communicating with its environment through wireless technologies such as Bluetooth, RFID, and Wi-Fi. Second, it must be capable of collecting data through sensors or actuators that gather information from its surroundings. These two capabilities allow a smart device to effectively monitor and respond to environmental changes, enhancing its utility and functionality. A prime example of a smart device is a smartwatch. When an ordinary watch is connected to a smartphone via Bluetooth, it can become a smartwatch capable of tracking various metrics such as calories burned, heart rate, steps taken, messages, weather forecasts, and sleep patterns, in addition to providing basic time and date functions. Smart devices generate vast amounts of data that can be stored in the cloud, processed, and analysed using advanced techniques such as genetic algorithms, neural networks, and AI. This analysis enables devices to make informed decisions. Transparency and reliable devices are essential to ensure data security and privacy [8].

As shown in Figure 8.2, IoT applications are transforming numerous industries, including smart homes, healthcare, smart cities, and agriculture. In smart homes, IoT allows users to control devices such as lights and appliances remotely, while in healthcare, it supports remote patient monitoring and wearable device tracking. IoT enhances traffic management, waste collection, and infrastructure maintenance in urban settings. In agriculture, it contributes to precision farming and livestock tracking. Manufacturing benefits from IoT through automation and optimised supply chains, and in transportation, it advances connected vehicles, fleet management, and smart parking. These examples highlight how IoT improves efficiency, convenience, and sustainability across sectors [9].

IoT has transformed numerous industries by enabling optimisation through data-driven insights. For example, automakers can use data from vehicle sensors to monitor and improve the efficiency and performance of their designs [10]. In addition to being able to access more data than would be possible using conventional data collection strategies, the real-time nature of IoT allows for immediate feedback. Although dealing with such large amounts of data can be problematic, decision-making can be better informed, and resources can be more effectively deployed thanks to the immediacy offered by IoT, which can also reduce waste and improve overall system performance.

IoT systems provide many benefits, but there are also important challenges and risks that arise due to (i) insecure/insufficient authentication and access controls; (ii) non-transparent/inadequate information usage consent and revocation mechanisms; (iii) shortcomings in the dependability of the IoT system service functionality; (iv) integration with other systems—(a) interaction of the IoT system with other

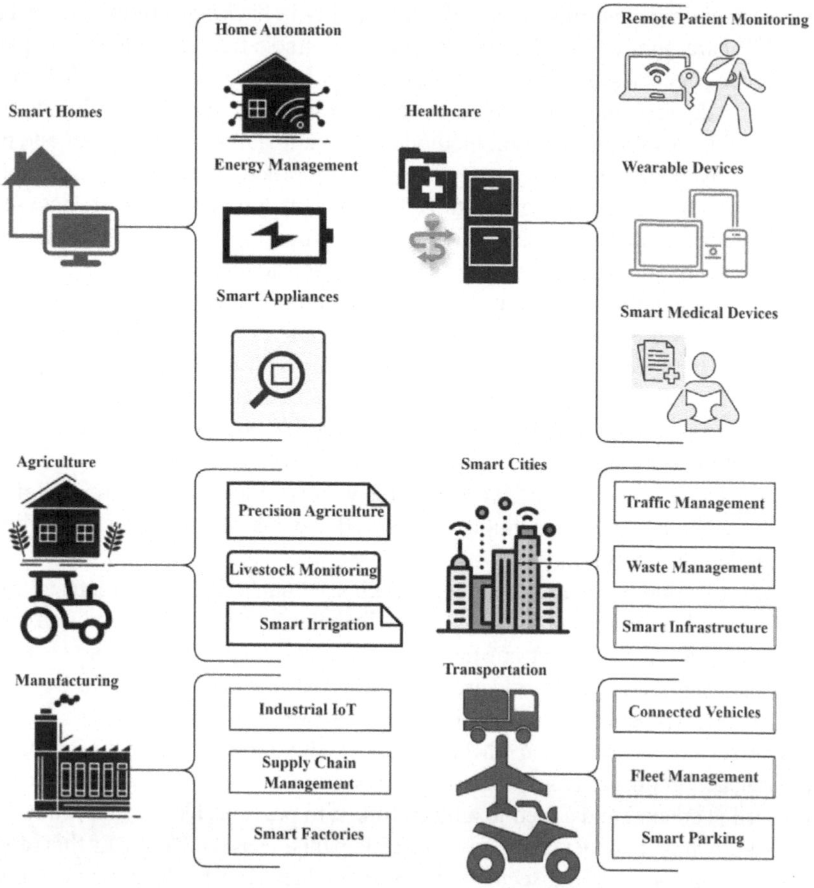

FIGURE 8.2 Applications of Internet of Things.

independently developed smart space applications, i.e., coexistence of independently managed objects in a common environment, and (b) interaction of other systems with IoT systems e.g., integration with existing legacy services into an IoT-enhanced enterprise/organisational IT infrastructure; (v) lifecycle issues—management of increasingly dynamic information including sensing, mining, and processing proprietary enterprise data resources/devices that do not have well defined or annotated public interfaces; (vi) their compliance to regulatory context-specific usage constraints, for example, in eHealth environments [11].

8.2.2 Blockchain

Blockchain is an Internet technology that uses the private key cryptography and incentive mechanisms to construct a decentralised, distributed, and shared immutable ledger of transaction blocks in chronological order, such that it ensures security in transactions [11]. On the contrary, issues exist with traditional bank systems, such as double spending, centralisation, and vulnerability to hacking activities, especially in the P2P paradigm of IoT. Against this context, innovation blockchain provides a highly secure decentralised solution for peer-to-peer transactions. In the IoT world, trust among devices and networks is extremely important. Blockchain builds trust by introducing novel authorisation methods to block unauthorised accesses and data manipulations. Its immutability makes altering recorded transactions challenging, enhancing security and making it suitable for payment systems. To ensure data security and integrity, blockchain uses elliptic curve cryptography and SHA-256 hashing algorithms to safeguard against unauthorised access and tampering [2].

Blockchain technology is a new alternative to traditional payment systems such as online wallets, banks, and third-party services. Let me explain this with an example: four friends (Jack, Ted, Sam, and Phil) decide to have dinner together and share the expenses. So, Phil can directly transfer money to Jack using an online payment service. At the same time, for Ted and Sam, it is impossible because of their bank issues, such as technical problems, account hacked issues, or transaction limits exceeding for the day. Cryptocurrencies will remedy this whole difficulty. They are digital currencies based on blockchain technology that works in a decentralised system without the mediation of banks [12].

Decentralisation has many advantages, such as enhanced security, reduced transaction fees, and increased control over users' financial activities [3]. Cryptocurrencies provide many advantages over normal money; they are impossible to counterfeit, do not rely on a central authority, and are protected by strong encryption [13]. Out of the thousands of cryptocurrencies, Bitcoin is the most well-known [14].

Cryptocurrencies are based on blockchain technology. Whenever a cryptocurrency transaction occurs (e.g., when one person sends money to another), a transaction record is created and added to a block linked to previous blocks in a chain. This chain is known as a ledger. All network participants keep their copy of this ledger, increasing transparency and security. Since the ledger is distributed across the network, it becomes difficult for hackers to manipulate information because others will see any modification to a transaction: their ledgers will not match. Data can also be hidden from unauthorised individuals using encryption algorithms.

Table 8.1 compares permissioned and permissionless blockchains. Permissioned blockchains restrict block creation for authorised nodes, require permissions for changes, and have high privacy. Instead, permissionless blockchains allow anyone to create blocks without changing permissions and having low privacy [4].

TABLE 8.1
Permissioned vs. Permissionless Blockchain

Feature	Permissioned Blockchain	Permissionless Blockchain
Node participation	Only authorised nodes can create blocks.	Anyone can create blocks.
Permission	Requires permission to make changes.	No permission is required to make changes.
Access	Typically, private.	Publicly accessible.
Privacy	High.	Low.

TABLE 8.2
Public vs. Private Blockchain

Feature	Public Blockchain	Private Blockchain
Permission	Permissionless	Permissioned
Access	Open to everyone	Restricted to authorised participants
Privacy	Low	High

Table 8.2 compares public and private blockchains with respect to their permissions, access, and privacy. A public blockchain is accessible to everyone, allowing anyone to participate in the network and view its transactions. Public blockchains are also permissionless (i.e., no special authorisation is required to create new blocks). A private blockchain is private for authorised users. Only one can create a block or access the network if they have permission to do so. This model can provide better privacy than public blockchains [4,15].

Bitcoin transactions use a dual-key system: a public key, which functions like an email address, and a private key, like a password. When sending Bitcoin, the user includes the recipient's address and amount and signs the transaction with their private key. A cryptographic algorithm generates a digital signature that verifies the sender's identity [7,16]. The signed transaction is then published on the Bitcoin network using the recipient's public key; however, only their private key can unlock and utilise the transaction. The first miner to solve the puzzle is rewarded with new Bitcoins. Blockchain is increasingly used across industries to improve security and transparency. For instance, Walmart employed blockchain to monitor product quality along its supply chain, identifying failures and ensuring better product outcomes. This illustrates blockchain's practical applications in solving real-world issues.

8.3 INTELLIGENT LEARNING MODELS

ILMs are advanced computing models that use learning techniques such as ML and DL algorithms to analyse large data sets. These models can uncover and learn intricate patterns and relationships that traditional analytics might not reveal. Also, the more varied the types of data ILMs can take in, the richer their insights or predictions will produce. Moreover, ILMs can adapt their performance over time as they learn from newer data, a defining feature in most IoT and blockchain use cases.

8.3.1 Artificial Intelligence

Table 8.3 summarises emerging trends and technologies in the blockchain-AI-IoT environment. It includes edge computing, federated learning, hybrid blockchains, secure hardware, data marketplaces, AI-powered blockchain analytics, self-sovereign identity, quantum-resistant cryptography, and energy-efficient blockchain solutions. These technologies offer various benefits, such as improved security, privacy, efficiency, scalability, and sustainability in blockchain-based systems.

TABLE 8.3
New Developments and Innovations in Blockchain: AI-IoT Integrated Ecosystems

Trend/Technology	Description	Benefits
	Edge Computing	
Edge AI	It combines the operation of edge computing with artificial intelligence for processing and decision-making on edge platforms.	Reduces latency, improves efficiency, and enables autonomous operations.
Edge blockchain	Applies blockchain on edge devices to support distributed consensus while storing data locally on the same devices.	Enhances security, privacy, and decentralisation in IoT and Industrial Internet of Things (IIoT) applications.
	Federated Learning	
Federated AI	Educates a number of edge devices with AI models while preserving the data.	Facilitates collaborative and decentralised AI training, protecting sensitive data.
	Interoperability and Standardisation	
Blockchain interoperability	Establishes protocols for seamless communication and data exchange between different blockchain networks.	Promotes collaboration and integration among blockchain-based systems.
IoT standardisation	Develops common standards and protocols for IoT devices and platforms.	Ensures interoperability and compatibility among IoT devices and applications.
	Hybrid Blockchain	
Public-private hybrid blockchain	Combines the benefits of public and private blockchains, offering scalability, privacy, and decentralised governance.	Provides a flexible and adaptable blockchain solution for IoT and IIoT applications.
	Secure Hardware	
Trusted execution environment	It provides secure hardware environments for conducting operations and protecting the storage of cryptographic keys.	Enhances the security of IoT and IIoT devices as well as their connectivity because of the segregation of the secure namespace and the shield of comprising services.
Hardware security modules	Provide safeguarded key storage and cryptographic functions.	Guarantees the security and privacy of information in IoT and IIoT implementations.
	Data Marketplace	
Blockchain-based data marketplace	Decentralised marketplaces for secure and transparent data exchange.	Enables individuals and organisations to monetise and trade their IoT and IIoT data.
	Blockchain Analytics Enhanced by Artificial Intelligence	
AI-driven blockchain analytics	Employs AI methods to examine blockchain data, identifying patterns, irregularities, and potential security risks.	Provides valuable insights and early warnings for potential issues in blockchain-based systems.

(*Continued*)

TABLE 8.3 (*Continued*)
New Developments and Innovations in Blockchain: AI-IoT Integrated Ecosystems

Trend/Technology	Description	Benefits
	Self-Sovereign Identity (SSI)	
SSI on blockchain	Leverages blockchain to enable individuals to have control over their digital identities.	Enhances privacy and security in IoT and IIoT ecosystems by giving individuals control over their data.
	Quantum-Resistant Cryptography	
Post-quantum cryptography	Explores algorithms resistant to quantum computing attacks.	Ensures long-term security for blockchain and IoT systems in the face of potential quantum threats.
	Energy Efficiency	
Green blockchain	Develop energy-efficient consensus algorithms and infrastructure.	Reduces the environmental impact of blockchain-based systems and promotes sustainability.

TABLE 8.4
Interdisciplinary Machine Learning, Blockchain, and IoT in Smart City Systems

Ref.	Topic	Summary
[17]	Convergence of ML, blockchain, and IoT	This work explores the potential of integrating ML, blockchain, and the IoT to create more advanced and collaborative smart city ecosystems.
[6]	ML applications in smart city IoT	As for this kind of smart city IoT system, this work is aimed at exploring the application of ML. It emphasises cases such as traffic flow, coordination of energy use, and outlier identification.
[18]	Blockchain for automation and data integrity	This work explores how blockchain technology and smart contracts can automate workflows and ensure data integrity across various smart city platforms.
[19]	Interoperability and data privacy challenges	This work addresses the challenges associated with interoperability between different systems and safeguarding data privacy within integrated smart city environments. Potential solutions are also discussed.
[20]	Federated learning and security enhancement	This paper aims to investigate how federated learning coupled with blockchain will improve security and privacy in the IoT context for smart cities.
[21]	Privacy-preserving frameworks	This work focuses on developing privacy-preserving frameworks that utilise Cloud and ML to ensure secure data key management.

8.3.2 Machine Learning

Table 8.4 provides a comprehensive overview of ML, blockchain, and IoT integration in smart city ecosystems. It explores the potential of these technologies to enhance various aspects of smart city development, including automation, data integrity, interoperability, data privacy, and security. The table highlights the challenges and

TABLE 8.5
Deep Learning-Based Models in IoT Applications for Smart Cities

Ref.	Methodology	Summary
[22]	Deep learning	ILMs are used in smart cities for governance, traffic management, healthcare, and crime prediction.
[23]	CNN-based	The implementation of convolutional neural networks (CNNs) for tasks like pedestrian detection, traffic signal control, crime prediction, and malware identification.
[24]	RNN-based	The application of recurrent neural networks (RNNs) for time-series forecasting, noise prediction, and small object detection in industrial environments.
[25]	Federated learning (FL)	FL is used for privacy-preserving data processing and decentralised analysis in smart city applications.
[12]	Deep reinforcement learning	Deep reinforcement learning optimises IoT systems such as traffic signals, autonomous vehicles, and smart energy grids in urban areas.
[26]	Deep and unsupervised learning	The deployment of deep and unsupervised learning for anomaly detection and pattern recognition in smart city IoT frameworks.
[27]	Transfer learning	Applying transfer learning in IoT systems for smart buildings, attack detection, and secure data fusion within industrial settings.
[28]	CNN and SVM-based	A CNN and Support Vector Machine (SVM)-based architecture for smart agriculture-based IoT systems.

solutions associated with integrating these technologies and discusses the potential benefits they can offer for smart city initiatives.

8.3.3 Deep Learning

An analysis of DL techniques and IoT applications in different fields has been made and used. DL techniques can learn and extract important features fundamentally, and IoT devices can forward real-time data to further DL models. In Table 8.5, we present a detailed analysis of DL models in IoT for sensing smart cities. It focuses on applying different types of DL methods, including central neural networks (CNNs), recurrent neural networks (RNNs), DRL, Deep Learning with Unsupervised Learning (DL-UL), and Transfer Learning (TL), to numerous purposes in smart city establishments.

8.4 OBSERVATIONS

Blockchain technology is certainly best suited for cryptocurrencies, but it has its uses in other fields. Thus, the fact of its activities in digital currencies became a valuable experience for its application in healthcare, economy, and governmental social benefits programs. Therefore, to unleash the potential of advanced technology, organisations need to gain a better understanding of the context and find use cases that can benefit from the elimination of intermediaries. However, the IoT has still not overcome security and privacy challenges because of its non-centralised structure and resource constraints, but it has become popular lately. Maybe that is why traditional security solutions cannot help secure IoT: millions of devices remain under a hacker's attack and open to data leaks.

It is, therefore, commensurate to focus on securing interaction between devices and their surroundings to enhance IoT security [29]. The combination with IoT, which is known as Blockchain-enabled Internet of Things (BIoT), can solve some of the problems present with the use of blockchain. The BIoT concept works with these technologies combined in order to lower the reliance on computationally intense mining algorithms, which could be more appropriate for lightweight IoT. To benefit fully from BIoT, a new framework needs to be defined in order to accommodate various distributed IoT applications [30]. Therefore, for BIoT to work, there is a need to address the scalability issue and reduce overhead traffic. The authors believe that blockchain database designers should strive to improve the speed of transactions and the number of transactions due to the computational burden that a large number of transactions place on IoT devices with limited power.

For the same reason that the internet exists, blockchain technology goes on changing and exists as well. Its IoT integration can create new standards of data openness and enhance interactions between similar devices. As these technologies progress, they could open many new markets of trillion-dollar worth.

8.5 FUTURE RESEARCH DIRECTIONS FOR BLOCKCHAIN-IOT INTEGRATION

Blockchain combined with IoT has certain issues that need further discussion to point directions to successful integration. Altogether, these challenges cannot be solved with one solution, but adequate and innovative solutions must be integrated.

Scalability: This is alarming, mainly given the amount of data on IoT, and the problem is worsening given the use of rapid 5G networks. Regarding this problem, some solutions can be used to enhance scalability. Sidechains and rollups, which are scaling solutions, are used above layer 2 to enhance throughput by reducing the load of the main blockchain through off-chain transactions. Sharding, the other approach, splits the main blockchain network into multiple sub-networks, thereby minimising traffic congestion and improving transaction rates. Also, other complex consensus algorithms, such as Proof of Stake and Delegated Proof of Stake, require fewer computational resources to back up scalability [22]. A combination of the features of public and private blockchains, known as hybrid blockchains, are characterised by their ability to have faster and more effective processes of transactions' execution while providing for data confidentiality and protection. In combination, these strategies appear to herald scalable handling of IoT within the constraints of a 5G setting.

8.5.1 Security and Privacy

These are some of the fundamental issues that are central to the use of blockchain and IoT. Future research in this area will focus on several perspectives. The PRIV-enhancing technologies are zero-knowledge proofs and homomorphic encryption. They provide means for protecting confidential information, and at the same time, they can be used to work with and transform in unsecured environments. Decentralised identity solutions ensure that users are in charge of their identity, so keeping your identity safe from fraudsters or hackers is not a big issue. Moreover, security created through smart contracts, encryption, and digital signatures can improve IoT devices or related transactions when based on blockchain technology. These advancements seek to produce a secure, privacy-preserving blockchain IoT environment [23].

Interoperability: This is a unique issue when it comes to IoT and blockchain integration because of the numerous protocols and standards implemented in both fields. Several important areas should be addressed in future research activities. First, the creation of standardised protocols is going to become a critical issue in the interactions between IoT devices and blockchain networks. It can also help create a single middleware framework that facilitates more efficient data exchange between IoT objects and blockchain platforms, eliminating incompatibility problems. Specific issues can also be addressed with reference to existing IEEE standards that include the use of blockchain with IoT: IEEE P2418.1, IEEE P2418.8, and IEEE P2418.9. It surfaced in an attempt to enhance the integration processes and make blockchain-powered IoT environments more integrated [24].

Edge Computing: This can, in turn, improve the performance and efficiency of blockchain-IoT systems since machine data processing will be done close to IoT devices. The above strategy brings the following benefits. First of all, the method minimises latency, which provides more opportunities for making decisions in real-time, and its importance is obvious for many applications of IoT. Second, edge computing promotes privacy and security due to handling sensitive data and not sending it online for processing. Finally, it aids in governance, where data handling reflects regulatory compliance at both regional and international levels. They make edge computing a useful element when it comes to improving and optimising blockchain IoT interfaces [25].

Federated Learning: This presents a useful way through which blockchain-IoT systems can be made even more effective since IoT devices can jointly train AI models locally. This approach results in two major advantages. First, it keeps data private, which are personal; the data do not go to central servers but remain on the users' devices. Second, federated learning provides enhanced efficiency because it avoids the transmitting and analysis of large volumes of data that are otherwise generative. This isolation of nodes suits the decentralised model well and adds to the security and functionality of IoT networks underpinning blockchain [25].

8.6 CONCLUSION

In this chapter, the combination of IoT and blockchain to build ILMs for optimising the performance and security of smart systems has been discussed. We began with the principles of IoT and how devices can convey and exchange particular information, as well as blockchain, which guarantees the integrity and authenticity of such shared information in a peer-to-peer network. More important is the discussion on the integration of the two technologies. The integration is referred to as BIoT, and this paper aims to explore it and look at its positives and negatives. We also explored how ILMs such as AI, ML, and DL can be used to analyse the large volumes of data created by IoT systems. These models contain superior analysis functions that support predictions on preventive maintenance, decision-making, and security. AI, ML, and DL, when incorporated into IoT environments, make the BIoT systems even more powerful because they have become smart systems. In the course of the chapter, we discussed some issues related to BIoT integration, including scalability, security, and privacy. We described several challenges within the context of actual applications, and we suggested potential solutions based on new consensus algorithms, anonymity solutions, and edge computing for decentralised data processing. Finally, we discussed research areas that are important for enhancing the use of IoT, blockchain, and ILM in the future. These are enhancing the applicability of blockchain for IoT for better scalability at large, enhancing privacy for the same by using enhanced cryptographic measures and analysing the use of federated learning to form a decentralised setup for the maintenance of AI. Mitigating these can fully unlock BIoT systems' opportunity for smarter, safer, and autonomous applications across various industries.

REFERENCES

1. Stankovic, J. A. Research directions for the Internet of Things. *IEEE Internet of Things Journal*, 1(1), 3–9, 2014.
2. Rao, A. R., Clarke, D. Perspectives on emerging directions in using IoT devices in blockchain applications. *Internet of Things*, 10, 100079, 2020.
3. Mu, X., Antwi-Afari, M. F. The Internet of Things (IoT) applications in industrial management: a science mapping review. *International Journal of Production Research*, 62(5), 1928–1952, 2024.
4. M. Wasid, R. Ali, S. Shahab. Adaptive genetic algorithm for user preference discovery in multi-criteria recommender systems. *Heliyon* 9(7), 2023.
5. Kizza, J. M. Internet of things (IoT): growth, challenges, and security. In *Guide to Computer Network Security*, pp. 557–573, 2024. Cham: Springer International Publishing.
6. Ansari, M. S., Wasid, M. Syed Atiqur Rahman. Devanagari handwritten character recognition using transfer learning with deep CNN and SVM. In *2022 5th International Conference on Multimedia, Signal Processing and Communication Technologies* (IMPACT), pp. 1–5. IEEE, 2022.
7. Shafik, W. Blockchain-based Internet of Things (B-IoT): challenges, solutions, opportunities, open research questions, and future trends. In *Blockchain-Based Internet of Things*, Reshi, I. A., Sholla, S. (eds.), pp. 35–58, 2024. Chapman and Hall.

8. Wasid, M., Anwar, K. Incorporating contextual information and feature fuzzification for effective personalized healthcare recommender system. In *Mathematical Modeling and Intelligent Control for Combating Pandemics*, pp. 197–211, 2023. Cham: Springer Nature Switzerland.
9. Ahmad, Shahnawaz, et al. Deep learning models for cloud, edge, fog, and IoT computing paradigms: Survey, recent advances, and future directions. *Computer Science Review,* 49, 100568, 2023.
10. Dorri, A., Steger, M., Kanhere, S. S., Jurdak, R. Blockchain: a distributed solution to automotive security and privacy. *IEEE Communications Magazine*, 55(12), 119–125, 2017.
11. Wasid, M., Anwar, K. An augmented similarity approach for improving collaborative filtering based recommender system. In *2022 International Conference on Data Analytics for Business and Industry (ICDABI)*, Sakhir, Bahrain, pp. 751–755, 2022. IEEE.
12. Wasid, M., Ali, R. A frequency count approach to multi-criteria recommender system based on criteria weighting using particle swarm optimisation. *Applied Soft Computing,* 112, 107782, 2021.
13. Makhdoom, I., Abolhasan, M., Abbas, H., Ni, W. Block chain's adoption in IoT: the challenges, and a way forward. *Journal of Network and Computer Applications*, 125, 251–279, 2019.
14. Rose, K., Eldridge, S., Chapin, L. The Internet of Things: an overview, understanding the issues and challenges of a more connected world, whitepaper released by internet society, 15 October 2015.
15. Dorri, A., Kanhere, S. S., Jurdak, R. Block chain in the Internet of Things: challenges and solutions, 2016, arXiv preprint arXiv:1608.05187.
16. Heidari, Arash, Nima Jafari Navimipour, and Mehmet Unal. Applications of ML/DL in the management of smart cities and societies based on new trends in information technologies: A systematic literature review. *Sustainable Cities and Society* 85, 104089, 2022.
17. Sha, M., Boukerche, A. Performance evaluation of CNN-based pedestrian detectors for autonomous vehicles. Ad Hoc Network. 128, 102784, 2022.
18. Rahhal, J. S., Abualnadi, D. IOT-based predictive maintenance using LSTM RNN estimator. In *Proceedings of the 2020 International Conference on Electrical, Communication, and Computer Engineering (ICECCE)*, Istanbul, Turkey, 12–13 June 2020, pp. 1–5, 2020. IEEE: Piscataway, NJ.
19. Zhang, C., Xie, Y., Bai, H., Yu, B., Li, W., Gao, Y. A survey on federated learning. *Knowledge-Based Systems*, 216, 106775, 2021.
20. Mushtaq, A., Haq, I. U., Imtiaz, M. U., Khan, A., and Shafiq, O. Traffic flow management of autonomous vehicles using deep reinforcement learning and smart rerouting. *IEEE Access*, 9, 51005-51019, 2021.
21. Wasid, M., Ali, R. An improved recommender system based on a multi-criteria clustering approach. *Procedia Computer Science*, 131, 93–101, 2018.
22. Gomez-Rosero, S., Capretz, M. A., Mir, S. Transfer learning by similarity-centred architecture evolution for multiple residential load forecasting. *Smart Cities*, 4, 217–240, 2021.
23. Khan, F. A., Khan, M. Z., Wasid, M., & Anwar, K. Soil classification and crop yield prediction using deep learning. In *2023 4th International Conference on Data Analytics for Business and Industry (ICDABI)* pp. 54–59, 2023. IEEE.
24. Ullah, Z., Naeem, M., Coronato, A., Ribino, P., De Pietro, G. Blockchain applications in sustainable smart cities. *Sustainable Cities and Society*, 97, 104697, 2023.
25 Azmi, M., Fathima, S. L., Wasid, M. Unveiling emotions through handwriting: a data analysis approach. In *2023 12th International Conference on Advanced Computing (ICoAC)*, Chennai, India, pp. 1–6, 2023. IEEE.

26. Masa'deh, R., Jaber, M., Sharabati, A. A. A., Nasereddin, A. Y., Marei, A. The blockchain effect on courier supply chains digitalization and its contribution to industry 4.0 within the circular economy. *Sustainability*, 16, 7218, 2024.
27. Hemashree, P., Kavitha, V., Mahalakshmi, S., Praveena, K., Tarunika, R. Machine learning approaches in blockchain technology-based IoT security: an investigation on current developments and open challenges. In *Blockchain Transformations: Navigating the Decentralised Protocols Era*, pp. 107–130, 2024. Springer.
28. Cui, L., Qu, Y., Xie, G., Zeng, D., Li, R., Shen, S., Yu, S. Security and privacy-enhanced federated learning for anomaly detection in IoT infrastructures. *IEEE Transactions on Industrial Informatics*, 18, 3492–3500, 2021.
29. Sezer, B. B., Turkmen, H., Nuriyev, U. PPFchain: A novel framework privacy-preserving blockchain-based federated learning method for sensor networks. Internet Things, 22, 100781, 2023.
30. Ahmad, Shahnawaz, Shabana Mehfuz, and Javed Beg. An efficient and secure key management with the extended convolutional neural network for intrusion detection in cloud storage. *Concurrency and Computation: Practice and Experience* 35(23), e7806, 2023.

9 Blockchain Innovation in IOT-Based Healthcare
Issues and Future Opportunities

Kedar Nath Singh, Arvind Kumar Upadhyay, and Arvind Mewada

9.1 INTRODUCTION

Internet of Things (IoT) devices, bulges, and systems must be able to collaborate despite their differences and provide various health services under Healthcare 4.0 to continuously check the health of patients [1]. Healthcare workflows incorporate services from several nodes. These include patient login information, medical consultations, real-time health record monitoring, and payments. All services are handled sequentially on platforms for cloud and fog computing [2]. The nodes are identical and exchange data according to a trust agreement. Hospital branches can share patient data, allowing convenient access anywhere. These hospitals also use services from payment processors, insurance providers, and others, which share secure data for processing. Blockchain has recently become a major technology with uses extending beyond Bitcoin. For example, the resource limits of IoT devices and the ledger-based structure of Blockchain protocols provide difficulties that necessitate future research. However, these advantages may expose potential attackers to security weaknesses and privacy issues.

The IoT is quickly increasing, with devices primarily used to monitor and control environmental changes, reduce dangers, and provide a wide range of useful services. Blockchain and IoT integration is still in its infancy because of resource limitations and ledger-centric architecture. Blockchain gives flexibility but also has the potential to generate privacy problems and security flaws that adversaries might exploit [3].

Blockchain technology has evolved dramatically since its inception, with each iteration bringing in innovations and discoveries, as shown in Table 9.1. The transition from cryptocurrencies to modern blockchain-based industrial applications represents a movement in using blockchain technology beyond digital currencies [4]. Blockchain applications are becoming increasingly popular in industries ranging from 1.0 to 10.0.

Blockchain technology to safely and dependably manage healthcare data is becoming increasingly popular. Blockchain revolutionises healthcare by facilitating

DOI: 10.1201/9781003545620-12

TABLE 9.1
Types of Blockchain Technology Required in Industry

Types of Blockchain	Application Description
Blockchain 1.0	Bitcoin, introduced in 2009, was the first implementation of blockchain technology. It focused on peer-to-peer electronic cash transactions. This allowed safe, decentralised digital payments without intermediaries like banks.
Blockchain 2.0	It is operated by Ethereum, which was founded in 2015. Smart contracts, or contracts that can be programmed to execute themselves, were introduced. Developers can construct both decentralised applications (DApps) and decentralised autonomous organisations. This expanded blockchain's applications beyond transactions. It now contains Decentralized Finance (DeFi), supply chain management, and other features.
Blockchain 3.0	It sought to address scalability difficulties that limited transaction throughput and speed. Sharing, sidechains, and layer two scaling (similar to Bitcoin's Lightning Network) are proposed. Interoperability standards promoted communication and data sharing between blockchain networks.
Blockchain 4.0	The purpose is to enhance the security, privacy, and governance mechanisms. New consensus algorithms have been launched, including Proof of Stake. Zero-knowledge proofs and privacy-focused cryptocurrencies like Zcash and Monero increased privacy. Governance systems have emerged to support decentralised decision-making and network enhancements.
Blockchain 5.0	Blockchain 5.0 intends to combine blockchain, AI, IoT, and big data. It emphasises sustainability and energy efficiency. Eco-friendly consensus procedures and renewable energy for mining are being developed. Enhanced regulatory compliance systems handle legal issues. These criteria promote mainstream adoption.
Blockchain 6.0	Blockchain 6.0 might put user experience first, making the technology more accessible. This involves improving user interfaces and minimising transaction prices. Improved scalability would allow for more users while maintaining speed. Blockchain-based reputation systems may enable less trustful interactions in various industries, including e-commerce and social networks. Environmentally friendly consensus processes should be prioritised. Energy-efficient protocols and carbon-neutral infrastructure can potentially lessen blockchain technology's environmental impact.
Blockchain 7.0	Scalability could overcome this issue through improved consensus mechanisms and sharing approaches. Other scalability options support more transactions per second. Standardised protocols and bridges enable seamless connectivity across various blockchain ecosystems. Advanced cryptography technologies enable users to transact without revealing sensitive information. Energy-efficient consensus methods can be implemented in the blockchain architecture.
Blockchain 8.0	Blockchain 8.0 may employ quantum-resistant cryptography to safeguard data against quantum attacks. Quantum computing poses a danger to present cryptography algorithms. It may allow users to establish and maintain reputations depending on their interactions. Immutable reputation systems may be introduced. These systems could be used to establish decentralised governance, verify identities, and foster trust. This would improve accountability and openness in online interactions.

(Continued)

TABLE 9.1 (*Continued*)
Types of Blockchain Technology Required in Industry

Types of Blockchain	Application Description
Blockchain 9.0	Blockchain 9.0, despite being theoretical, has the potential to transform blockchain technology. It could include advancements in numerous sectors. Future technological breakthroughs will decide its specific features. The blockchain community's efforts, as well as societal requirements, will shape its potential.
Blockchain 10.0	It might fully integrate quantum computing capabilities. This encompasses quantum-resistant cryptography and quantum consensus techniques. These features would provide unsurpassed security and performance.

secure data sharing and enhancing diagnosis and treatment efficiency. By integrating all real-time clinical data about a patient's health into an already-existing, safe healthcare framework, blockchain has the potential to play a significant role in the future provision of tailored, genuine, and secure healthcare. This study uses blockchain technology to look at recent and upcoming developments in the health sector business [5]. A blockchain network has been utilised by healthcare providers to confidentially store and share patient data between pharmacies, doctors' offices, hospitals, and diagnostic labs. Blockchain-based applications can identify serious and potentially harmful faults in healthcare.

Consequently, it enhances the healthcare system's medical data interchange's efficiency, safety, and transparency. Medical organisations benefit from this technology since it provides valuable information and improves the analysis of medical records [6]. Blockchain technology provides a distributed, secure database that may function without management or central authority. Blockchain continuously creates a growing list of consecutive items called blocks via a distributed, peer-to-peer network combined to build a digital ledger. Every transaction is included in a cryptographically signed block that is automatically verified by the network. Due to its potential to enhance data transparency and authenticity through various applications, including limiting access to electronic health records (EHRs) and streamlining the claims process, blockchain is gaining popularity in the healthcare sector [7]. A robust and productive healthcare system depends on interoperability since it enables software and technical platforms to use and share data between various health organisations and application providers securely and efficiently. The healthcare sector is currently dealing with problems including fragmented and compartmentalised data, delayed communications, and disjointed workflow tools due to a lack of interoperability. Blockchain provides a solution by facilitating the access of longitudinal, tamper-evident, and comprehensive medical records stored in disparate systems while preserving pseudo-anonymity and security [8].

Blockchain technology is a distributed, transparent, and safe numerical transaction scheme that uses blocks to store data, with transactions being the fundamental units. Consensus techniques ensure data security and integrity through cryptographic hashing. Nodes, computers, or servers participate in the peer-to-peer network design, allowing direct communication without a central authority. The details of the primary elements of blockchain technology are provided in Table 9.2.

TABLE 9.2
Primary Components of Blockchain Technology

Primary Components	Components Description
Distributed ledger	A digital ledger distributed across numerous computers or nodes. Each participant received an identical copy. This ensures transparency while also lowering the danger of fraud.
Cryptography	Used to protect data and maintain transaction integrity. Public and private keys allow for safe data transfer and user authentication. Hashing creates a unique digital fingerprint for each block, assuring data integrity.
Consensus mechanism	Nodes utilise protocols to determine transaction validity and blockchain state. Proof of Work and Proof of Stake are two common consensus techniques. Other options are Practical Byzantine Fault Tolerance and Delegated Proof of Stake.
Blocks	Data structures store transaction information. Each block contains a header containing metadata. This includes a date, a hash value of the preceding block, and a nonce (in PoW systems). The block also includes transaction details.
Smart contracts	Contracts that execute themselves and have their terms written in code. These contracts automatically enforce obligations depending on predetermined criteria. They remove the need for intermediaries.
Decentralisation	The network runs on a peer-to-peer basis, with no central authority. Each node possesses equal authority. This approach is censorship-resistant and avoids centralised points of failure.
Immutability	The information is recorded in a block and uploaded to the blockchain, and it is unchangeable or unavoidable. Cryptographic hashing and block linking ensure this. Tampering with previous records becomes incredibly tough.
Transparency and anonymity	Transactions are transparent and accessible to anyone on the network. The participants' real names are concealed. Public addresses replace personal information. This improves privacy while remaining transparent.

A digital ledger is crucial for financial services and operations, and blockchain technology ensures data encryption and reduces data breaches. It allows for multiple shared copies of a single database, reducing the risk of attacks. Blockchain is transforming the banking industry by efficiently tracking and documenting tangible and intangible assets [9]. Consensus methods, access control, and purpose categorise blockchain technology. The network consists of nodes, each with a shared ledger, updated regularly and subject to verification processes. The classification of blockchain technology is depicted in Figure 9.1.

9.2 LITERATURE SURVEY

A deep review of the blockchain technology is shown in Table 9.3.

9.3 IMPORTANT IOT-BASED HEALTHCARE ISSUES

IoT-based healthcare provides numerous advantages, including enhanced patient monitoring, data collection, and operational efficiency, but it also has several significant drawbacks. Healthcare systems heavily trust data collected by IoT sensors,

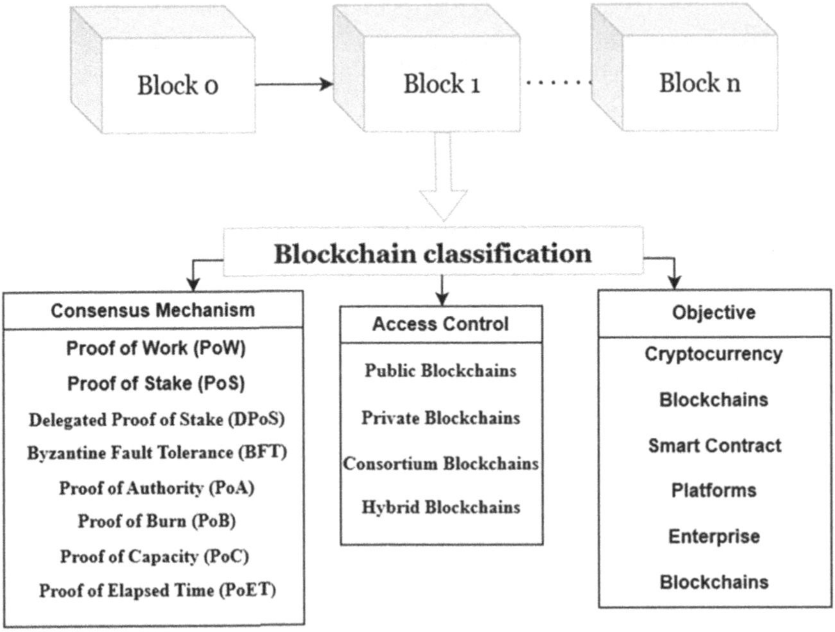

FIGURE 9.1 Classification of blockchain technology.

TABLE 9.3
Comparative Analysis of Literature

Authors	Methodology	Literature Review	Limitations
[23]	This paper comprehensively reviews how blockchain technology impacts IoT security and highlights the obstacles blocking its wider adoption in the IoT.	This synopsis gives an extensive grasp of IoT design, security, and performance needs.	To determine the obstacles to blockchain adoption in the IoT context, the study examined blockchain principles, IoT requirements, and blockchain-based applications.
[24]	An overview of blockchain integration with IoT for decentralised systems is given in the text, along with a discussion of the advantages, difficulties, and potential future study areas.	IBM ADEPT project integrates blockchain with distributed devices, while Slock. It offers an Ethereum platform for IoT integration.	Blockchain scalability issues lead to centralisation, impacting cryptocurrencies' future. IoT systems' centralised server/client architecture also presents scalability and authentication issues.

(Continued)

TABLE 9.3 (*Continued*)
Comparative Analysis of Literature

Authors	Methodology	Literature Review	Limitations
[25]	The protocol for Leased Proof of Stake (LPoS) consensus, along with consensus processes and smart contracts, is a key component of thin-layer blockchain technology for Internet of Things applications.	The proposed low-power blockchain technology for IoT smart home applications aims to improve privacy protection and enable device customisation and control using SC blockchain.	Consensus systems are vulnerable to attacks such as the 51% attack, and Proof of Work protocol issues like centralisation, latency, and power consumption can cause problems.
[26]	The techniques employed are SVM, automation, robotics, Artificial Neural Network (ANN), Decision Tree (DT), Natural Language Processing (NLP), and AI.	A comprehensive insight into smart IIoT, blockchain, and AI technologies, addressing tools, strategies, security, unresolved issues, promising developments, and potential taxonomies while addressing privacy and confidentiality challenges demonstrated.	Resource limitations and security issues are hindering the implementation of Blockchain in Internet of Things devices.
[27]	IoT solutions in healthcare utilise RSA encryption for secure communication, distributed ledgers, smart contracts, and consensus methods for execution and validation.	The article discusses consensus methods and Rivest–Shamir–Adleman (RSA) encryption for data protection while highlighting the significance of blockchain and IoT tiers in data security.	The blockchain-based intrusion detection system addresses security flaws, including interoperability, scalability, and regulatory compliance.
[26]	Blockchain systems utilise the SHA-256 algorithm for cryptographic hash functions, utilising optimisation algorithms for system performance improvement and safe data verification in digital twins.	This paper explores the potential of blockchain technology, digital twins, and the IoT in pandemic control, suggesting a decentralised, secure digital asset replication infrastructure.	The high cost of digital twin applications, concerns about data security, privacy, latency, and moral dilemmas pose significant challenges for businesses.
[27]	Applied AMOS software for confirmatory factor analysis and construct validation, using rat scales to measure constructs ranging from 1 to 7.	This model is re-examined, highlighting additional obstacles and variables and critiquing previous research for overlooking factors contributing to innovation rejection.	This paper includes cultural disparities, budgetary constraints, and experimental technique limitations.

(*Continued*)

TABLE 9.3 (*Continued*)
Comparative Analysis of Literature

Authors	Methodology	Literature Review	Limitations
[28]	IoT and blockchain applications are used in healthcare systems, focusing on medical record management, remote patient monitoring, and medication traceability.	The paper thoroughly examined the integration of IoT and blockchain in the healthcare sector, utilising various internet resources.	This paper focuses on security, privacy, standards, interoperability, laws, regulations, and development.
[29]	The paper assesses the security of blockchain-based IoT-based smart healthcare systems through proximity security techniques, behaviour detection, and encryption.	A comparative analysis and description of the use of blockchain technology in smart healthcare systems.	Cryptographic techniques may face computational challenges, while blockchain's latency and throughput limitations for medical applications and security and privacy issues in current EHR systems hinder adoption.
[30]	The solution utilises clinical sharing for nearby clinics and hospitals, EHR sharing, Medchain, distributed blockchain data protection, and dual encryption.	This paper explores blockchain technology's principles, goals, limitations, and performance, its future applications in IoT, the evaluation of innovative initiatives, and its application in the medical field.	The proposed schemes have not undergone performance testing, indicating insufficient review of the associated access control work.
[31]	Statistical analysis and machine learning techniques are utilised for data interpretation, combining data from various sources for a comprehensive analysis.	To explore the integration of AI, ML, and data analytics in the healthcare sector.	This paper's small sample size may result in potential bias due to self-reported data.
[32]	Blockchain technology in IoT applications, as well as theoretical research and simulation, is used to analyse the properties of BBP-based databases.	The article emphasises the importance of considering BBP-based databases in the IoT context, highlighting potential issues for further research in the IoT ecosystem.	Blockchain-based IoT databases in dynamic networks face issues with consistency, platforms, and outcomes, and the use of BBP for distributed key-value databases has fundamental constraints.
[33]	To explore blockchain technology's potential in managing medical data, focusing on its advantages and disadvantages.	Blockchain technology's advantages and potential challenges in healthcare data management highlight its potential in the healthcare sector.	The paper discusses the obstacles to implementing blockchain technology in healthcare data management systems.

which must be protected from unauthorised modifications due to security and privacy concerns.

Privacy and Data Security: IoT applications in healthcare can boost system effectiveness, promote patient well-being, and facilitate quick emergency reactions. Data processing poses significant privacy security risks, as well as documentation, monitoring, and data transfer of these systems. Patients' lives may be in danger if treatment and diagnosis are delayed due to worries about security, privacy, and healthcare data protection. IoT implementation improves patient outcomes, increases system performance, and speeds emergency response [10]. However, there are substantial threats to the security and privacy of data when it comes to these systems throughout processing, documentation, monitoring, and transfer. Problems with the confidentiality, privacy, and protection of medical data may put patients in danger by delaying diagnosis and treatment. Healthcare IoT technologies can significantly enhance emergency response times, system performance, and patient care. However, they also present significant privacy and security issues regarding the processing, documentation, monitoring, and data transfer. These worries about privacy, security, and data protection may cause delays in diagnosis and treatment, which could endanger patients' lives [11].

Cooperation: Collaborative features in IoT-based healthcare systems must be implemented in a decentralised environment using a distributed algorithm that ensures efficient resource sharing and equitable distribution of rewards among involved users. The simulation results indicate that this approach consistently improves system-wide and user-specific performance. In addition, the strategy ensures fair distribution of resources [12].

The Ability to Scale Devices: Recent advancements in computing and smart device technology have led to the generation of massive amounts of "Big Data" that are crucial for forecasting future trends. However, some technologies struggle to handle this complexity, causing concerns and issues that require attention [13]. Healthcare providers require scalable infrastructure to manage this inflow of data without sacrificing efficiency.

Integration and Management of Data: IoT network for e-health systems, aiming to improve security, safety, and data confidentiality. It suggests collecting data from smart health devices, and filtering and analysing it for efficient and cost-effective diagnosis and treatment of various illnesses. Using smart fitness and health devices connected to the IoT blockchain network to collect and analyse health data enables efficient and cost-effective diagnosis and treatment for various illnesses and ensures data confidentiality.

Commitment to Regulations: The IoT is revolutionising various sectors, including healthcare, by enabling real-time data collection and analysis despite numerous technological challenges. Healthcare IoT solutions must adhere to laws regulating patient privacy and data protection, which can be challenging to navigate and require constant attention [14].

User Training and Adoption: Exploring the use of IoT-based wearable devices in ageing healthcare records and examining the effects of factors promoting and inhibiting their use. Wearables with IoT technology offer real-time health monitoring, increased independence, and improved communication with healthcare professionals, enabling quick responses to abnormalities. Older individuals may face challenges in adopting wearables with IoT capabilities, such as data security, privacy, cost, and stigma concerns. They may fear becoming dependent on digital devices [13]. Healthcare workers need to get the proper training to use IoT devices and analyse the data they provide.

Network Connectivity: IoT-based healthcare systems require stable network connectivity for real-time data, especially for wearable monitors and smart sensors. Disruptions can lead to data loss or delays, so backup connectivity options are crucial for emergencies. Encrypting data transmitted over networks, implementing strong authentication and authorisation mechanisms, and preventing unauthorised access to sensitive health data are crucial measures to protect patient confidentiality. The growing number of IoT devices in healthcare necessitates a scalable network infrastructure to handle the significant data volume generated by these devices without compromising performance. Inadequate network infrastructure might hinder the adoption and efficacy of IoT-based healthcare solutions, especially in rural or disadvantaged areas.

Ethical and Legal Issues: IoT technology, including smart devices and applications, collects and monitors data, resulting in big data. Protecting this data from cyber threats requires advanced encryption techniques, secure communication protocols, and robust authentication systems. Ethical and legal concerns arise from collecting and using big data, as personal information can lead to privacy violations. Regulations such as **General Data Protection Regulation** (GDPR) and **Health Insurance Portability and Accountability Act** (HIPAA) provide guidelines for enforcement, but this is complicated due to the global nature of the Internet and IoT. To overcome these challenges, coordinated efforts from technology developers, policymakers, and international bodies are needed to ensure IoT technology's safe and ethical use [14].

Battery Life and Power Management: The IoT comprises sensor nodes that monitor physiological signals like **Electroencephalogram** (EEG), **Electrocardiogram** (ECG), blood pressure, and temperature using cellular networks, Wi-Fi, Bluetooth, etc. In the health sector, the Internet of Medical Things consists of resource-constrained wearable devices. The foremost task is energy drain and battery consumption in these used devices. Numerous IoT devices are battery-operated, and it can be inconvenient to constantly change or recharge batteries, particularly for wearable or implanted devices [15].

9.4 BLOCKCHAIN TECHNOLOGY SOLVED INTERNET OF THINGS-RELATED HEALTH PROBLEMS

Blockchain technology can completely transform IoT-based healthcare systems, offering revolutionary possibilities for IoT-enabled healthcare. Blockchain technology and IoT can significantly improve healthcare services, innovate, and drastically alter IoT-enabled healthcare systems.

9.4.1 Improved Privacy and Security of Data

The IoT has led to the development of distributed ledgers and blockchains, offering enhanced security, anonymity, and capacity. However, these technologies face limited storage, high resource use, transaction rate scalability, predictability, and legal issues. The latest approaches aim to resolve these issues by integrating blockchain technology with IoT. Despite these challenges, innovative solutions are being explored to merge blockchain technology with IoT to improve security and privacy in IoT-driven healthcare systems. Blockchain has the potential to revolutionise IoT-based healthcare solutions [16]. The decentralised and unchangeable ledger of blockchain guarantees the security and imperviousness of health data transferred and stored via IoT devices. To protect their privacy and ensure that only authorised parties see their data, patients can securely and openly give and withdraw access to their health records.

9.4.2 Enhanced Interoperability and Data Integrity

Advancements in information technology and blockchain technology have transformed the healthcare industry, enabling miniaturised sensors for monitoring vital signs and portable health devices for self-health tracking. However, inconsistent security policies and access control structures hinder these advancements. Blockchain-based platform for monitoring patient vital signs, utilising Hyperledger Fabric and Libelium e-Health toolkit. This innovative approach provides patients with an immutable history log and global access to medical information, enhancing data security and user comfort in medical sharing [17]. Blockchain enhances medical record integrity by creating an immutable patient data history, reducing errors due to corrupted or altered data, and ensuring accurate diagnosis and treatment. Blockchain technology enhances clinical trial credibility by securely collecting and verifying data from IoT devices, reducing the risk of manipulation and fraud.

9.4.3 Effective Clinical Trials and Investigations

Blockchain technology enables a transparent, traceable record of the entire supply chain, enhancing the management of medical supplies and pharmaceutical authenticity. IoT devices in clinical trials collect patients' health data in real-time, utilising blockchain technology for secure recording, efficient monitoring, and prompt protocol modifications. The detailed taxonomy of blockchain technology in clinical trials highlights its potential for comprehensive, secure, and transparent coverage. It also addresses open research challenges in technical, security, and organisational aspects. Blockchain technology can securely and transparently handle clinical trials but faces technical, security, and organisational challenges [18]. Blockchain technology enables real-time health data collection from clinical trial patients, enhancing efficiency and facilitating quicker protocol adjustments.

9.4.4 Supply Chain Management

Integrating blockchain with the IoT can develop a smart health supply chain management system. In addition to preventing forged medications and harm to medical components reported, this strategy ensures authentication, lowers costs, and monitors the status of products during their shipment from manufacturers to end-users. It also offers decentralised tracking and tracing of medicinal items, privacy, security, trust, and visibility. A smart supply chain requires a secure infrastructure to ensure the timely and accurate delivery of goods to the right person. At the same time, IoT devices collect real-time health data for supply chain network control, but incomplete information and conflicting stakeholder behaviour make tracking medical components challenging [19]. Blockchain ensures patient safety by preventing the entry of counterfeit medications into the market by tracking every stage of the supply chain on an immutable ledger.

9.4.5 Automated Smart Contracts

Insurance claim handling can be automated with smart contracts, which lowers administrative costs and expedites claim settlements. This can be especially helpful in managing the vast amounts of information that IoT healthcare equipment generates. The smart contract is a section of programming that is added to a blockchain and starts running automatically when certain requirements are satisfied. Blockchain is a distributed, mutual record where exchanges are meticulously documented and linked to each other to give the whole provenance or history of the benefit. Smart contracts apply to a number of businesses, such as financial services and banking, wherever they enable numerous investors to offer finances to multiple borrowers at the same terms. Smart contracts are used in the energy industry to distribute energy generally. The Digital Right Management System allows the entertainment sector to handle movie and music rights more easily through the use of smart contracts [20]. Insurance claim handling can be automated with smart contracts, which lowers administrative costs and expedites claim settlements. This can be especially helpful in managing the vast amounts of information that IoT healthcare equipment generates. Automatic payments for medical services are made under predetermined guidelines. Smart contracts have the ability to initiate payments, for example, when a patient's IoT device signals that a health check-up is finished.

9.4.6 Improved Telemedicine Patient Monitoring

Blockchain can help IoT devices monitor patients remotely by guaranteeing that the data collected is reliable and unchangeable. This is necessary to give patients trustworthy care outside of conventional hospital settings. Medical IoT devices collect bio-signals for personalised healthcare, enhancing individual health-related data management. The Individual Health Records System ensures privacy, individual control, data integration, safety, and flexibility. Blockchain technology is emerging as a potential selection aimed at managing health information. The healthcare industry is transitioning from a hospital-centric to a person-centric approach, promoting national engagement in disease prevention and early detection [21]. Thanks to blockchain technology, paramedics and emergency services can instantly access a patient's medical information in an emergency, which can be essential for prompt and efficient treatment.

9.4.7 Decentralised Health Information

Healthcare organisations are adopting innovative technologies to improve patient and provider control over health information. Blockchain technology offers a decentralised structure for smooth international services and enables medical professionals to record and oversee patient care without a central authority. However, privacy concerns and competitive edge concerns prevent institutional adoption. Cloud-based systems and blockchain technology enhance record security, providing patients with a dedicated area for organising and safeguarding their health records and allowing access at any time and from multiple facilities [22]. Blockchain enables the construction of decentralised medical records that people may take to other institutions and healthcare providers. This guarantees full, up-to-date medical histories and continuity of service.

9.5 CONCLUSION

Blockchain technology, originally developed for Bitcoin, has gained considerable recognition since its inception. This chapter examines several kinds of blockchain technology and their uses in various fields, including IoT-based healthcare. Applications in the music business, financial services, IoT, edge computing, smart grids, cybersecurity, and healthcare have all expressed great interest in it due to its potential. Blockchain integration improves interoperability, security, and efficiency in IoT-based healthcare systems, propelling digital transformation and bettering patients' health. Patients always wish to protect their medical records and not give them to any healthcare organisation, whether through phones or other technological means. Today's research on IoT-based healthcare offers a novel way to employ IoT resources that are underutilised safely and securely using blockchain technology.

REFERENCES

1. H. Kayan, M. Nunes, O. Rana, P. Burnap, C. Perera, "Cybersecurity of Industrial Cyber-Physical Systems: A Review," *ACM Computing Surveys*, 54(11s), 1–35, 2022.
2. D. Karthikeyan, B. Amutha, K. Mekata Devi, N. Singh, P. Anglin Kuruba, "Hybrid Rater to Quantify and Measure the Severity of Infection and Spread of Infection in Muskmelon," *Bulletin of Electrical Engineering and Informatics (BEEI)*, 13(3), 1667–1675, 2024. https://doi.org/10.11591/eei.v13i3.5432.
3. M. A. Ansari, D. K. Singh, "Human Detection Techniques for Real Time Surveillance: A Comprehensive Survey," *Multimedia Tools and Applications*, 80, 8759–8808, 2021. https://doi.org/10.1007/s11042-020-10103-4.
4. Z. Wenhua, F. Qamar, T.-A.N. Abdali, R. Hassan, S. T. A. Jafri, Q. N. Nguyen, "Blockchain Technology: Security Issues, Healthcare Applications, Challenges and Future Trends," *Electronics*, 12, 546, 2023. https://doi.org/10.3390/electronics12030546.
5. R. K. Dewang, M. P. Yadav, S. Awasthi, O. Raj, A. Mewada, K. L. Bawankule, "Data Secure Application: An Application That Allows Developers to Store User Data Securely Using Blockchain and IPFS," *Multimedia Tools and Applications*, 83(15), 45491–45517, 2024.

6. N. Goenkaa, A. K. Sharmab, S. Tiwaric, N. Singhd, V. Yadave, S. Prabhuf, K. Chadaga, "A Regularized Volumetric ConvNet Based Alzheimer Detection Using T1-Weighted MRI Images," *Cogent Engineering*, 11(1), 2314872, 2024. https://doi.org/10.1080/23311916.2024.2314872.
7. N. Singh, S. P. Sasirekha, A. Dhakne, B. V. Sai Thrinath, D. Ramya, R. Thiagarajan, "IOT Enabled Hybrid Model with Learning Ability for E-Health Care Systems," *Measurement: Sensors*, 24, 1–14, 2022. https://doi.org/10.1016/j.measen.2022.100567.
8. P. Paul, P. S. Aithal, R. Saavedra, S. Ghosh, "Blockchain Technology and Its Types—A Short Review," *International Journal of Applied Science and Engineering (IJASE)*, 9(2), 189–200, 2021.
9. M. Tiwari, N. Singh, A. Mewada, M. A. Ansari, "Machine Learning Empowered Breast Cancer Diagnosis: Insights from Coimbra Dataset Analysis," *Recent Advances in Computer Science and Communications*, 11, e26662558297605, 2024. https://doi.org/10.2174/0126662558297605240926055723.
10. P. K. Bishoyi, S. Misra, N. Kumar, "Collaborative and Efficient Body-to-Body Networks for IoT-Based Healthcare Systems," *IEEE Internet of Things Journal*, 9(14), 12147–12154, 2022. https://doi.org/10.1109/JIOT.2021.3133794.
11. N. Singh, Y. Kumar, "Security and Privacy Requirements in 5G Healthcare," In *Block Chain for 5G Healthcare Applications*. IET Publication, pp. 1–30, 2021. https://doi.org/10.1049/PBHE035E_ch1.
12. T. Frikha, A. Chaari, F. Chaabane, O. Cheikhrouhou, A. Zaguia, "Healthcare and Fitness Data Management Using the IoT-Based Blockchain Platform," *Journal of Healthcare Engineering*, 2021, 9978863, 2021.
13. J. Calvillo-Arbizu, I. Román-Martínez, J. Reina-Tosina, "Internet of Things in Health: Requirements, Issues, and Gaps", *Computer Methods and Programs in Biomedicine*, 208, 106231, 2021. https://doi.org/10.1016/j.cmpb.2021.106231.
14. B. Sivathanu, "Adoption of Internet of things (IoT) Based Wearables for the Healthcare of Older Adults—A Behavioural Reasoning Theory (BRT) Approach," *Journal of Enabling Technologies*, 12(4), 169–185, 2018. https://doi.org/10.1108/JET-12-2017-0048.
15. N. Singh, H. P. Singh, A. Mishra, A. Khare, M. Swarnkar, S. K. Almas, "Blockchain Cloud Computing: Comparative Study on DDoS, MITM and SQL Injection Attack," In *2024 IEEE International Conference on Big Data & Machine Learning (ICBDML)*, Bhopal, India, 2024, pp. 73–78. https://doi.org/10.1109/ICBDML60909.2024.10577412.
16. H. Magsi, A. H. Sodhro, M. S. Al-Rakhami, N. Zahid, S. Pirbhulal, L. Wang, "Novel Adaptive Battery-Aware Algorithm for Data Transmission in IoT-Based Healthcare Applications," *Electronics*, 10(4), 367, 2021. https://doi.org/10.3390/electronics10040367.
17. H. D. Zubaydi, P. Varga, S. Molná, "Leveraging Blockchain Technology for Ensuring Security and Privacy Aspects in Internet of Things: A Systematic Literature Review," *Sensors*, 23, 788, 2023. https://doi.org/10.3390/s23020788.
18. F. Jamil, S. Ahmad, N. Iqbal, D.-H. Kim, "Towards a Remote Monitoring of Patient Vital Signs Based on IoT-Based Blockchain Integrity Management Platforms in Smart Hospitals," *Sensors*, 20(8), 2195, 2020. https://doi.org/10.3390/s20082195.
19. L. Hang, C. Chen, L. Zhang, J. Yang, *Blockchain for Applications of Clinical Trials: Taxonomy, Challenges, and Future Directions*. IET Communications, 2022. https://doi.org/10.1049/cmu2.12488.
20. S. K. Nanda, S. K. Panda, M. Dash, "The Medical Supply Chain Is Integrated with Blockchain and IoT to Track the Logistics of Medical Products," *Multimedia Tools and Applications*, 82, 32917–32939, 2023.
21. D. Negi, A. Sah, S. Rawat, T. Choudhury, A. Khanna, *Block Chain Platforms and Smart Contracts*. EAI/Springer Innovations in Communication and Computing (EAISICC). Springer, 2021. https://doi.org/10.1007/978-3-030-65691-1_5.

22. P. Pawar, N. Parolia, S. Shinde, T. O. Edoh, M. Singh, "eHealthChain—A Blockchain-Based Personal Health Information Management System," *Annals of Telecommunications*, 77, 33–45, 2022.
23. G. Atlas, K. P. Arjun, Bindu Babu, *A Decentralised Privacy-Preserving Blockchain for IoT and Big Data in Healthcare Applications.* CRC Press, 2021. https://doi.org/10.1201/9781003081180-2.
24. A. O. Almagrabi, R. Ali, D. Alghazzawi, A. AlBarakati, T. Khurshaid, "Blockchain-as-a-Utility for Next-Generation Healthcare Internet of Things," *Computers, Materials and Continua*, 68(1), 359–376, 2021. https://doi.org/10.32604/cmc.2021.014753.
25. J. Almalki, "State-of-the-Art Research in Blockchain of Things for HealthCare," *Arabian Journal for Science and Engineering*, 49(3), 3163–3191, 2024. https://doi.org/10.1007/s13369-023-07896-5.
26. H. F. Atlam, A. Alenezi, M. O. Alassafi, G. B. Wills, "Blockchain with Internet of Things: Benefits, Challenges, and Future Directions," *International Journal of Intelligent Systems and Applications*, 10(6), 40–48, 2018. https://doi.org/10.5815/ijisa.2018.06.05.
27. E. El-Din Hemdan, W. El-Shafai, A. Sayed, "Integrating Digital Twins with IoT-Based Blockchain," *Wireless Personal Communications*, 131, 2193–2216, 2023.
28. M. M. Kamruzzaman, B. Yan, M. N. I. Sarker, O. Alruwaili, M. Wu, I. Alrashdi, "Blockchain and Fog Computing in IoT-Driven Healthcare Services for Smart Cities," *Journal of Healthcare Engineering*, 2022, 9957888, 2022.
29. I. Makhdoom, M. Abolhasan, W. Ni, "Blockchain for IoT: The Challenges and a Way Forward," In *ICETE 2018—Proceedings of the 15th International Joint Conference on e-Business and Telecommunications*, vol. 2, pp, 428–439, 2018. https://doi.org/10.5220/0006905604280439.
30. Z. Mani, I. Chouk, "Consumer Resistance to Innovation in Services: Challenges and Barriers in the Internet of Things Era," *Journal of Product Innovation Management*, 35(5), 780–807, 2018. https://doi.org/10.1111/jpim.12463.
31. P. S. Pavithra, P. Durgadevi, "Improving Security: Blockchain Based IoT Solutions for the Healthcare," *Journal of Theoretical and Applied Information Technology*, 102(6), 2716–2725, 2024.
32. A. D. Preetha, T. S. P. Kumar, "Securing IoT-Based Healthcare Systems from Counterfeit Medicine Penetration Using Blockchain," *Applied Nanoscience*, 13(2), 1263–1275, 2023. https://doi.org/10.1007/s13204-021-01984-4.
33. A. Rahman, D. Kundu, T. Debnath, M. Rahman, A. A. Aishi, J. Islam, "Blockchain-Based AI Rahman Methods for Managing Industrial IoT: Recent Developments, Integration Challenges and Opportunities," https://arxiv.org/abs/2405.12550.

10 Blockchain-Driven Supply Chain Management in Industry 5.0

Enhancing Efficiency and Sustainability

Harsh Raj, Garima Jain, Sanat Jain, Prafull Farswal, Vartika Singh, and Rajasekaran Selvaraju

10.1 INTRODUCTION

The objective of the forthcoming fifth industrial revolution, or Industry 5.0, is to integrate human creativity with the accuracy, velocity, and intelligence of robotics. It is reinstating individuals to the manufacturing floor. Potential advantages of Industry 5.0 include advancements in cloud manufacturing, intelligent healthcare, manufacturing production, and supply chain management. There are various supporting approaches that help in the functioning of Industry 5.0. Blockchain technology is one of them [1].

Industry 5.0 has been derived from Industry 4.0. The shift from Industry 4.0 to Industry 5.0 is not too complex, but blockchain technology is seen to play a vital role in various sectors. In Industry 4.0, blockchain technology was used to make supply chains more secure and transparent [2]. In Industry 5.0, this is taken to the next level. For example, it facilitates the assurance of high-quality, genuine products. It allows companies to place custom orders through special apps, and transactions are managed using smart contracts. This makes things such as payments and logistics much smoother and cost-effective. In manufacturing, blockchain plays a massive role. From procuring raw materials and using automated processes in manufacturing to delivering the finished product, blockchain helps ensure that operations are efficient, secure, and transparent [3]. It aids in storing records of students in such a manner that it is secure, hard to modify by unauthorised sources, and easy to share.

Furthermore, it caters to innovative experiences such as remote labs and holographic teaching. For the Unmanned Aerial Vehicle (UAV)/drone operations and disaster management domains, blockchain will keep sensitive data safe while

 DOI: 10.1201/9781003545620-13

TABLE 10.1
Transition from Industry 4.0 and Industry 5.0 [4]

Aspect	Industry 4.0 (Human-Out-of-the-Loop)	Industry 5.0 (Human-Back-into-the-Loop)
Focus	Technology-driven: Automation, smart factories, cyber-physical systems, and machine autonomy	Value-driven: Human-centricity, sustainability, resilience, and well-being of workers and society
Role of human	Humans are largely removed from industrial processes; systems and robots handle most tasks autonomously.	Humans play a significant role in collaboration with robots and systems, focusing on human–machine cooperation.
Technology	Dominated by AI, IoT, cloud computing, machine learning, and cognitive computing.	Integrates weak AI for human control, focusing on transparent human–machine cooperation.
Communication	Machine-to-machine communication	Human–machine interaction with collaborative robots
Decision-making	Autonomous decision-making by machines	Humans participate in critical decision-making alongside machines
Goal	Efficiency and productivity, data-driven automation	Human-centric, sustainable, and resilient industrial processes that align with societal and ecological needs
Impact on jobs	Some job loss due to automation and autonomous systems	Industry aims to create new jobs by incorporating humans into more collaborative, innovative, and decision-making processes.
Ethical considerations	Largely ignored or secondary, "Black Box AI" with little human control	Focus on ethical decisions and fostering transparent human–machine cooperation, especially during system failures or degraded modes.

ensuring on-time and secure communications in drone networks or vehicle systems. Blockchain is, in simple words, one of the most imperative tools moving into Industry 5.0 to make our industries more secure, efficient, and connected. Below is Table 10.1, which uses various parameters to differentiate between Industry 4.0 and Industry 5.0. Think of blockchain as a super-secure, worldwide digital ledger. This is immutable, unlike regular databases, and lots of people can see it. The result of this is that an organisation cannot control it. Thus, if you make a transaction, then the system will check that, and everybody can see what happened. In light of the emergence of Industry 5.0 and blockchain technology, this study aims to achieve the following objectives:

RO1: To analyse the principal benefits of blockchain within the supply chain management domain in Industry 5.0.
RO2: To identify and assess the essential enablers of blockchain adoption for supply chain management in Industry 5.0.

10.2 LITERATURE REVIEW

Innovations or processes enhancing efficiency through blockchain technologies or processes that augment the efficiency of systems for supply chain management in Industry 5.0 and the role of blockchain therein.

Achieving this involves several technologies and processes, with the Internet of Things (IoT) being particularly significant. The IoT constitutes an extensive network of intelligent devices—gadgets, vehicles, and structures—all interconnected via the Internet. These gadgets possess electronics, software, and sensors that enable data collection and sharing. This link facilitates remote monitoring and control, enhancing overall efficiency. Both real and virtual entities within the IoT ecosystem have unique characteristics that allow them to interconnect and become part of a larger information network. Various technologies, including barcodes, wireless sensor networks, smart sensing, Near Field Communication (NFC), Radio Frequency Identification (RFID), low-energy cloud computing, and wireless communications, play essential roles in IoT [5].

The IoT is revolutionising different industries by bringing smart applications like smart homes, cities, self-driving cars, and wearable devices. IoT's role in the supply chain system is to improve inventory management through technologies like RFID and wireless networks for a real-time tracking and automation system. Industry 4.0 brings new information into inventory models—a movement from supplier-driven production to customer-demand-driven. It would involve time-varying production rates as well as policies regarding inventory replenishments in a manner that fully enables the warehouse to prepare and set up for changed demand in a highly effective manner [6].

Cooperation is the new gold standard for human-smart system relationships in the age of Industry 5.0, which includes robots. When it comes to improving this partnership, the IoT is king. Envision a smart factory where machines share data about manufacturing processes and make choices in real time through IoT communication. This connectivity boosts efficiency because machines can work together seamlessly. It also improves the operations that take place in the warehouse, where one can calculate the fastest route, which, in turn, speeds up product delivery and improves inventory management. Industry 5.0 places more emphasis on a balanced interaction between humans and machines, as compared with Industry 4.0, which aims to minimise human involvement. This version allows humans to take up creative roles while robots handle repetitive tasks. The combination of both further improves production quality. Industry 5.0 uses the IoT in such a way that industries transform into a collaborative environment, where combining human creativity with the capabilities of robots creates more agile and scalable operations [7].

The IoT network is much more strengthened and secure by combining new cutting-edge technology with it. To better understand the implications of blockchain, think of it as an extra safeguard that prohibits unauthorised parties from intercepting data transmissions between smart home devices. Blockchain technology enables simple and highly secure communication between various devices, like thermometers, refrigerators, and webcams. Distributed ledger technology (blockchain) makes it far more difficult to corrupt or manipulate the system than storing data in a single vulnerable area [8]. Secure digital commitments made across several devices enable smart contracts to facilitate interactions between parties. Many vulnerabilities exist in traditional security systems because they rely on a single authority. Contracts with intelligence, however, are reliable, independently executed, and impenetrable. They ensure that only authorised devices have access to the correct information, thereby maintaining the integrity of both the data and the overall IoT system. Therefore, smart contracts on blockchain enhance IoT security without requiring a complete overhaul of existing systems [9]. This technology also empowers IoT devices to process transactions, making it incredibly useful for autonomous machines that need to make quick, secure payments for services [10].

It is evident from the above that blockchain records each entry of data collected from IoT on a distributed ledger, thus verifying the integrity of IoT data. This transparent recording, making it tamper-proof, will increase the level of reliance among stakeholders. Predictive analytics can rely on this secure data to not only give forecasts in demand patterns but also detect anomalies in operation, such as failures of machinery or failure of the supply chain and allocate resources optimally. Integrate blockchain's security with the analytical capabilities of artificial intelligence (AI) to unlock a whole new avenue of making more data-driven decisions by businesses with more confidence for improving both the efficiency and the ethical practices of global supply chains [11].

Information sharing is critical for businesses in supply chain systems to thrive in the context of Industry 5.0. Technological advancements have made it easier to share data across different levels of the chain, enhancing coordination and cooperation among suppliers, producers, retailers, and distributors. Without sufficient information sharing, organisations cannot operate effectively [12]. It can increase a company's competitive advantage and improve responsiveness to market changes while ensuring better alignment of products with consumer demand. Although sharing information has associated costs, the benefits, particularly through advancements in Enterprise Resource Planning (ERP) and web-based systems, are significant. Therefore, data sharing in the supply chain is a key factor in achieving better coordination, integration, and efficiency throughout the entire supply chain. Blockchain technology enforces trust and secures shared data by storing information in encrypted form.

It gives data owners complete control over who can access their information and under what conditions. The transparency and immutability of blockchain records also instil trust among users, making it a powerful solution for industries dealing with high volumes of sensitive data [13]. Thus, blockchain emerges as a vital component for ensuring secure transactions and protecting personal and sensitive information. Traditional methods of verifying the authenticity of shared documents often require time-consuming offline processes involving contacting issuing authorities. The dispersed and irreversible characteristics of blockchain computing resolve this issue by offering a visible, collective ledger of activities that can promptly authenticate the validity of shared information. Blockchain's cryptographic strength prevents unauthorised entities from editing records, further enhancing security [14].

Blockchain ensures that ethical sourcing is maintained in a global supply chain by providing transparent and immutable records of every step in the sourcing and production process. Regarding the certifications for ethically sourced materials, blockchain authenticates the certifications that comply with fair trade or standards of sustainable practices. Traceability gives a confident assessment of the origins of products both by consumers and companies, holding suppliers accountable to ethical and environmental standards. This will create a better, responsible, and transparent supply chain ecosystem because blockchain eliminates and prevents fraud and risks of unethical conduct within the supply chain [15].

Because of the mentioned features, blockchain computing is characterised as a transformational solution for different sectors, especially for the health and energy sectors. Specifically, in the health sector, mainly in oncology, blockchain can significantly enhance data sharing with security, personalised care, and medical research through the features of transparency, traceability, and immutability. However, it should be noted that no amount of technology can protect data privacy or security and can be merged with other cryptographic methods. Legal, social, and technological barriers to adoption have kept large-scale real-world deployment at bay thus far. However, future research can overcome all of these through the development of privacy-preserving systems that will be interoperable with other systems and that are compliant with international regulations [16].

In the energy area, blockchain is driving change from centralised to decentralised energy systems, focusing currently on smart grids, microgrids, and peer-to-peer energy markets. China has led the basic blockchain research for energy; however, most real-world applications arise in developed countries. Future direction and research and development in this sector should be directed at integrating the two sectors so that basic research and practical applications are bridged [17].

Emerging trends such as quantum blockchain and the addition of AI in blockchain are going to revolutionise Industry 5.0, with both aspects related to decentralised systems and human-centric production. Quantum AI can greatly scale up efficiencies in blockchain by speeding up transaction validation and maintaining data safety while coupling AI algorithms with the computational power that comes from quantum computing. This integration will ensure efficient healthcare, financial services, and supply chain management through better prediction and analysis, as well as fraud detection. These technologies will look towards improving automation and decision-making to make industrial solutions sustainable and intelligent when applied in Industry 5.0. For this context, the combination of blockchain systems aligns with the requirement of Industry 5.0, which aims to improve the way humans and machines work together, as well as make manufacturing efficient. Figure 10.1 shows the framework of the proposed methodology.

10.3 ISSUES INDUSTRIES MAY FACE IN INDUSTRY 5.0

Security: In this context, increasing digitalisation and connectivity through the IoT increases the risk of cyberattacks against factories and industrial processes. This has made security issues a top priority in Industry 5.0. Such attacks can cause humongous financial losses. Security breaches have the potential to disrupt important industrial productions, delivery, and customer satisfaction timelines. To overcome this, secure identity management,

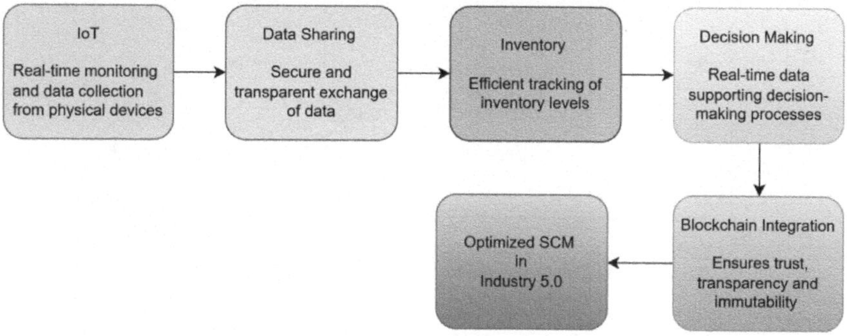

FIGURE 10.1 Proposed framework.

constant system monitoring, and advanced encryption are essential security attributes. To improve the reliability of digital ecosystems and prevent unauthorised access, businesses can use blockchain technology in conjunction with the IoT to securely exchange data with one another [18].

Privacy: Industry 5.0 aims to meet human needs through customised product production, relying heavily on user data to achieve such personalisation. Consequently, securing user data is vital. If user data falls into unauthorised hands, it can be exploited for identity theft, financial fraud, or unauthorised surveillance. As data sharing increases among multiple parties, controlling access becomes increasingly complex. A lack of control can easily lead to misuse, highlighting the need for a privacy-by-design approach, where privacy is embedded in the technology and business processes to strengthen overall system security. The unique feature of blockchain lies in its decentralisation, which grants individuals control over their data. By limiting access to certain users in accordance with permissions specified in intelligent agreements, blockchain technology helps participants more effectively handle their personal information [18].

Data Sharing: Despite its importance in Industry 5.0, data sharing does have some drawbacks. Complicating factors in information sharing include a lack of trust between organisations and different policies and standards. As a result, it is critical to establish an ecosystem that promotes data sharing among different players. Data ownership concerns and the fear of competitors gaining access to sensitive information can discourage organisations from sharing their data.

Centralisation: This parameter appears to be one major obstacle that could stop Industry 5.0 from achieving its goal. This exposes them to cyber-attacks and malfunctions. The compromise of centralised structures could lead to serious consequences not just for the organisation but for everyone who is a part of the supply chain. Also, flexibility and response are two terms that are crucial requirements for Industry 5.0, where consumer expectations and marketplace circumstances are always changing. Figure 10.2 shows the flow of process.

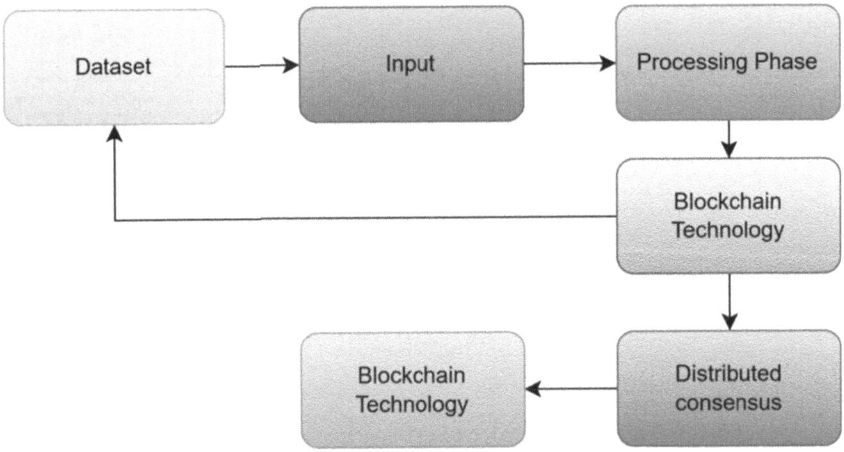

FIGURE 10.2 Proposed flow structure.

10.4 ROLE OF BLOCKCHAIN IN ADDRESSING THE CHALLENGES OF INDUSTRY 5.0

It is important to address the potential challenges that all the organisations that are a part of some supply chain may encounter as Industry 5.0 develops. Blockchain technology is key to solving these problems. If exploited correctly, it has great potential to facilitate the safe integration of new technology, thereby fostering digital transformation in Industry 5.0. The decentralised, transparent, and secure nature of blockchain technology makes it ideal for solving many problems that are seething in Industry 5.0, including the issues that are related to data sharing, privacy, and centralisation. For example, blockchain improves reliability by spreading data that has to be stored across a decentralised network, eliminating single points of failure. Crucial for preserving the authenticity of the supply chain, the invincibility of the data on a blockchain guarantees that information remains tamper-proof.

The use of effective encryption techniques and access controls that are based on smart contracts further enhances blockchain's privacy features. It gives users more control over their personal information by allowing them to set specific permission for data access. This control helps build trust between participants, allowing organisations and people on a personal level to share information while protecting their privacy. Blockchain removes common issues that occur while working together by securely sharing data between organisations. The fact that it could hold records permanently and also provide a shared view makes it easier for stakeholders to trust each other. This is particularly helpful in handling supply chains because many people rely on accurate data to make decisions.

Blockchain and gamification blend in a perfect synergy, which could be drawn on the supply chains. The stakeholders involved in the supply chain can be motivated to collaborate to increase involvement and transparency using blockchain-based tokens or rewards. For instance, participants receive tokens once they have met their particular performance markers or in case of timely deliveries, and the blockchain will handle secure, immutable records of their progress. This thereby integrates gamification through blockchain: it motivates stakeholders and strengthens accountability and trust by reducing the problem of transparency to decentralised supply chains [19].

Blockchain technology enables every user to have control over their own data in the chain and contribute to data sovereignty. The decentralised and open nature of the ledger used in blockchain ensures that data is not controlled by a single entity but distributed across a highly secure network. This allows individuals to manage access to their private data through the use of smart contracts, where they can determine the type of people allowed to access or use their information and the conditions under which such people should access them. Data is also cryptographically locked down to trace the movement along the supply chain while allowing users to maintain their privacy. This allows users to trust that personal data can be accessed only by authorised parties, thereby improving transparency without taking away user control.

The availability of the decentralised nature of blockchain will give individuals all sovereignty over data, and by doing this, it will minimise its reliance on third parties that were central to the risks of misuse of data [20,21]. Blockchain provides privacy because of the transparent and trustworthy chain of information. Figure 10.3 shows the research in Industry 5.0.

The stored data in the blockchain are secure, and third-party access is only possible with proper permissions and, thus, completely private [22]. One of its excellent features is its transparency and reliability, which makes it an apt platform for shared data across several systems. This offers secure, efficient data exchange in business, government, and logistics systems. Hence, it acts as a very trustable method for any real-time or other kind of data-sharing process. Blockchain is a technology through which peer-to-peer networks can be created. Through these remarkably diverse production units, machines, devices, or even a single manufacturer, they can be connected without any intermediaries or centralised control. It is further in a position to provide end-to-end transparency across the supply chain and thus allow

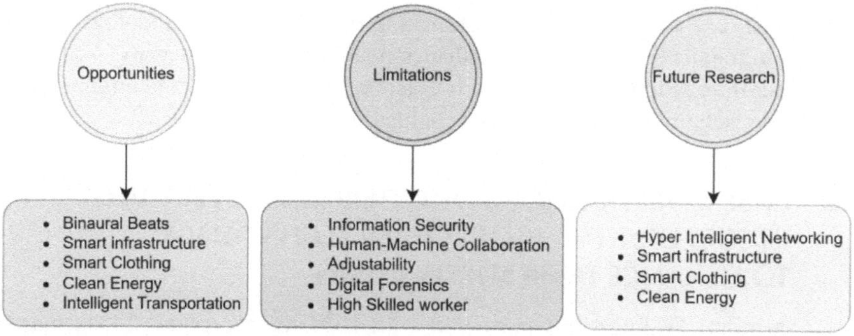

FIGURE 10.3 Analysis of opportunities, constraints, and prospects for future research in industry 5.0.

production units to trace raw materials, components, and finished products. This way, transparency gives each unit involved in the production process a clear view of the whole supply chain, thus enhancing coordination and trust [23]. It also provides real-time inventory tracking, where firms can trace the flow of merchandise within the supply chain. This transparency ensures that levels of inventory are current and accurate, thereby averting situations wherein businesses have a stockout or overstock situation [24].

10.5 ISSUES WITH THE IMPLICATION OF BLOCKCHAIN TECHNOLOGY

One of the key drawbacks of blockchain technology, for instance, is high energy consumption in Proof of Work models. In this model, miners must rely on solving extremely resource-hungry cryptographical puzzles to reach a consensus. As a result, electricity usage is quite high and provokes more environment-related concerns. The network, of course, cannot scale indefinitely in its ability to deal with the number of transactions since the amount of time and computational power necessary to validate will increase. Other limitations include regulatory challenges and interoperability issues among the many blockchain networks that exist in the world today [25]. Further, the integration of blockchain technology with legacy systems also poses a big challenge, including architectural incompatibility due to the centralised nature of legacy systems versus the decentralised nature of blockchain. Data synchronisation is complex, as legacy systems often use different formats that must be aligned with the blockchain's structure. The cost and complexity of modifying outdated infrastructure can be prohibitive, while scalability and transaction speed limitations of blockchain may not meet the demands of high-volume legacy systems. Concerns regarding security and privacy emerge from blockchain's transparency, potentially conflicting with the privacy standards of legacy systems [26].

To address the difficulty of seamless interoperability, standardisation of data exchange protocols is necessary to facilitate successful information transfer among diverse platforms. Legacy systems can connect to the blockchain using Application Programming Interfaces (APIs), thereby reconciling architectural discrepancies to facilitate data exchange in a safe environment. A decentralised methodology, coupled with cryptographic protections, can enhance data privacy and integrity. Addressing stakeholder integration and scalability is essential for management to attain efficient and safe blockchain interoperability throughout the network [27].

10.6 CASE INVESTIGATIONS OF ENTERPRISES EFFECTIVELY INCORPORATING BLOCKCHAIN TECHNOLOGY TO ENHANCE THEIR SUPPLY CHAIN

1. **Walmart:** Walmart has partnered with IBM to incorporate blockchain technology into its supply chain to enhance the monitoring of food goods' provenance. Using IBM's Food Trust blockchain, Walmart can trace a

product in seconds, from the farm to the store shelf, improving transparency, security, and productivity. As a result, companies may respond more rapidly to food recalls, which helps to identify and remove any tainted products swiftly [8].

2. **De Beers (Tracr):** To monitor the movement of diamonds across their supply chain, De Beers, the diamond corporation, introduced a blockchain platform named Tracr. This promotes the ethical sourcing of diamonds and helps stop the circulation of diamonds that have been in conflict (often called "blood diamonds"). According to Ahmed and Abdurrahman [9], blockchain technology enables more openness in monitoring the whole lifecycle of a diamond, from mining to storage.
3. **Maersk (TradeLens):** With the help of IBM's TradeLens platform, Maersk—one of the world's leading shipping companies—has integrated blockchain technology. By digitalising and simplifying logistics and shipping paperwork, this technology enables efficient worldwide trade. Zhan and Wang [20] state that TradeLens employs blockchain technology to provide real-time shipping transaction updates, which reduces paperwork, enhances supply chain transparency, and minimises fraud.

10.7 CONCLUSION

Industry 5.0 signifies the latest industrial revolution centred on the collaboration between human ingenuity and machine productivity. It was founded on Industry 4.0, with the incorporation of blockchain technology to enhance security and transparency in supply chains. Industry 5.0 utilises blockchain technology to guarantee the authenticity and superior quality of products. It also works on a distributed ledger system, which is used to remove middlemen such as banks by creating a secure record of transactions. When it comes to the IoT, data interchange, inventory management, and decision-making processes, this technology is vital for many different types of enterprises. Here, the term "Internet of Things" refers to a system of interconnected Internet-enabled smart devices. It is important in Industry 5.0 because it allows remote monitoring and control. Integrating blockchain with IoT would enhance both reliability and security across these systems, allowing secure communication between sensors. Blockchain also ensures secure data sharing through encryption and access control, which is vital for efficient supply chain management.

The handling of inventory will become more efficient using blockchain as it transforms supply chain logistics. Furthermore, the platform's transparency and immutability facilitate trustless collaboration, which in turn benefits the decision-making process. The primary component of blockchain in the new era of Industry is the integration of the IoT with the standards for interoperability techniques such as Proof of Stake and Proof of Work. An essential need in Industry 5.0, given the massive amounts of data produced by IoT devices, this integration guarantees the security of information, integrity, and overall automated transactions via smart contracts among devices, encouraging smooth and secure operations.

REFERENCES

1. Maddikunta, P. K., Pham, Q.-V., B, P., Deepa, N., Dev, K., Gadekallu, T. R., Ruby, R., and Liyanage, M. (2022a). Industry 5.0: A survey on enabling technologies and potential applications. *Journal of Industrial Information Integration,* 26, 100257. https://doi.org/10.1016/j.jii.2021.100257
2. Gokhale, P., Bhat, O., and Bhat, S. (2018). Introduction to IOT. *Computer Networks, 5*(1), 41–44.
3. Javaid, M., Haleem, A., Singh, R. P., Rab, S., and Suman, R. (2021). Significance of additive manufacturing for Industry 4.0 implementation. *Automation*, 6(2), 33.
4. Fatima, Z., Tanveer, M. H., Waseemullah, Zardari, S., Naz, L. F., Khadim, H., Ahmed, N., and Tahir, M. (2022). Production plant and warehouse automation with IOT and Industry 5.0. *Applied Sciences, 12*(4), 2053. https://doi.org/10.3390/app12042053
5. Fosso Wamba, S., Guthrie, C., Queiroz, M. M., and Minner, S. (2024). ChatGPT and generative artificial intelligence: an exploratory study of key benefits and challenges in operations and supply chain management. International Journal of Production Research, 62(16), 5676–5696.
6. Lone, A. H., and Naaz, R. (2021). Applicability of blockchain smart contracts in securing internet and IOT: A systematic literature review. *Computer Science Review, 39*, 100360. https://doi.org/10.1016/j.cosrev.2020.100360
7. Mathur, S., Kalla, A., Gür, G., Bohra, M. K., and Liyanage, M. (2023). A survey on the role of Blockchain for IoT: Applications and technical aspects. *Computer Networks, 227*, 109726. https://doi.org/10.1016/j.comnet.2023.109726.
8. Bahsi, H., and Akleylek, S. (Eds.). (2023). AI-enabled threat detection and security analysis for cyber-physical systems. Springer. https://doi.org/10.1007/979-8-8688-0315-4
9. Jussen, Ilka and Schweihoff, Julia and Dahms, Valentin and Möller, Frederik and Otto, Boris. (2023). Data Sharing Fundamentals: Definition and Characteristics.
10. Hugos, M. H. (2024). Essentials of supply chain management. John Wiley and Sons.
11. Akinsulire, A. A., Idemudia, C., Okwandu, A. C., and Iwuanyanwu, O. (2024). Supply chain management and operational efficiency in affordable housing: An integrated review. Magna Scientia Advanced Research and Reviews, 11(2), 105–118.
12. Winkelbauer, L. (2023). *Blockchain for ethical supply chain sourcing.* https://www.larswinkelbauer.com/blockchain-for-ethical-supply-chain-sourcing/
13. Makaram, P., and Gheorghiu, R. (2021). Applications of blockchain technology for data-sharing in oncology: Results from a systematic literature review. *Oncology Research and Treatment*, 98(6), 403–411. https://doi.org/10.1159/000514799
14. Kouhizadeh, M., and Sarkis, J. (2021). Blockchain practices, potentials, and perspectives in greening supply chains. *IFAC PapersOnLine*, 54(1), 81–86. https://doi.org/10.1016/j.ifacol.2021.04.002
15. Adel, A. Future of Industry 5.0 in society: human-centric solutions, challenges, and prospective research areas. J Cloud Comp 11, 40 (2022). https://doi.org/10.1186/s13677-022-00314-5
16. Haris, M., and Yasin, S. (2023). Integrating blockchain technology into knowledge management: A comprehensive review. *Global Knowledge, Memory and Communication.* https://doi.org/10.1108/GKMC-11-2023-0426
17. Paiva, J. M., Alves, J. M., and Lima, J. M. (2024). The role of blockchain in promoting sustainability: A systematic literature review. *Sustainability*, 16(7), 2823. https://doi.org/10.3390/su16072823
18. Abdullah, N., Hakansson, A., and Moradian, E. (2017). Blockchain-based approach to enhance big data authentication in a distributed environment. *2017 Ninth International Conference on Ubiquitous and Future Networks (ICUFN).* https://doi.org/10.1109/icufn.2017.7993927

19. Wen, B., Wang, Y., Ding, Y., Zheng, H., Qin, B., and Yang, C. (2023). Security and privacy protection technologies in securing blockchain applications. *Information Sciences*, *645*, 119322. https://doi.org/10.1016/j.ins.2023.119322
20. Centobelli, P., Cerchione, R., Vecchio, P. D., Oropallo, E., and Secundo, G. (2022). Blockchain technology for Bridging Trust, traceability, and transparency in Circular Supply Chain. *Information and Management*, *59*(7), 103508. https://doi.org/10.1016/j.im.2021.103508
21. Tripathi, G., Ahad, M. A., and Casalino, G. (2023). A comprehensive review of Blockchain technology: Underlying principles and historical background with future challenges. *Decision Analytics Journal*, 9, 100344. https://doi.org/10.1016/j.dajour.2023.100344
22. Shukla, A., and Soni, P. (2021). Blockchain technology: Limitations and future possibilities. *ResearchGate*. https://doi.org/10.4018/978-1-7998-2414-5.ch009
23. Vartak, P., Nandan, R., and Kamble, S. S. (2023). A systematic review of blockchain technology for supply chain management. *Information*, 11(11), 509. https://doi.org/10.3390/info11110509
24. Shamsudin, N., and Ali, M. A. (2022). Exploring the impact of blockchain technology on supply chain transparency: A systematic literature review. *The European Proceedings of Social and Behavioural Sciences*, 2022, 58. https://doi.org/10.15405/epms.2022.10.58
25. Razali, M. N., Rahman, A., and Zainuddin, M. (2019). The role of blockchain in promoting transparency in the supply chain. In A. F. A. Rahman, M. N. Razali, and M. Zainuddin (Eds.), *Advances in Business and Management* (pp. 189–199). Springer. https://doi.org/10.1007/978-3-030-05764-0_18
26. Ahmed, R., and Abdurrahman, A. (2021). Blockchain-based solutions for supply chain transparency: A review. *Automation*, 5(3), 43.
27. Zhan, Y., and Wang, L. (2021). The impact of blockchain technology on the performance of supply chain: Evidence from China. *Management Supply Chain Review*, 25(2), 203–221. https://doi.org/10.1108/MSCRA-10-2020-0028

Section 4

Societal and Industrial Impacts

11 Healthcare Security and Privacy Concerns in the Blockchain Network

A Review

KM Manisha and J Sathish Kumar

11.1 INTRODUCTION

In Machine Learning (ML)/Deep Learning (DL)-based healthcare applications, there is an absence of actual medical datasets. Doctors and healthcare providers find it challenging to disclose sensitive data for the purpose of developing strong and efficient machine learning and deep learning models that can quickly and accurately diagnose diseases, which is why these medical datasets remain private. In light of the exponential growth of healthcare data and the growing desire for data-driven medical innovations, the ethical and safe sharing of private patient data is a critical concern. A fine balance must be struck by healthcare organisations, researchers, and patients between the critical requirement to protect patient privacy and data security and the usefulness of aggregated medical data for scientific advancement.

The problem has sparked the development of cutting-edge technologies that have the potential to change the way that healthcare data is shared completely. When multiple medical systems are used, there is a risk to patient information security, privacy, and accessibility. If security is compromised, there is also a risk to availability, data integrity, and privacy. Hackers can use medical records to find prescriptions and commit fraud or identity theft by ordering medications online. They can also use extortion and blackmail to force victims to declare private medical information.

Finally, there is a risk to the mobility, privacy, and security of IoMT-enabled health tools such as fitness trackers [1]. Single points of failure, interoperability problems, and concerns about data privacy and security are some of the drawbacks of centralised data-sharing systems. Decentralised data sharing, which disperses data among numerous nodes or peers without the need for a central authority or middleman, has emerged as a viable solution to these problems. Decentralised data sharing also has a number of advantages, including better data ownership and control, higher privacy and security, and increased responsibility and openness [2].

Blockchain: Blockchain is an immutable, decentralised, append-only public ledger. The blockchain is a public ledger, as anyone can participate in the transaction, and every node will keep a copy of it and maintain its copy.

DOI: 10.1201/9781003545620-15

Everybody acts as a node and processes data locally; there is no central database to hold the chain's data. This is called decentralisation [3]. Block is a container data structure that contains a series of transactions. Each block is linked together through its hash values. By using cryptographic hashes to link blocks together, a blockchain's integrity is preserved while guaranteeing decentralisation and tamper resistance. A block is divided into two parts—block header and block body. Block headers contain version, timestamp, hash, difficulty level, nonce, and Merkle root. Versions define the version of the block structure. The amount of time to create a block is called a timestamp. A hash function takes input of any length and produces fixed-length output.

The difficulty level indicates the level of complexity and processing power needed to mine the network. Nonce is a counter used for the proof of work algorithm [4]. The Merkle root serves as a blockchain's unique identifier for each transaction within a block. The block body contains the list of transactions. Larger blocks can help in processing a large number of transactions in one go. However, verification and propagation take longer. Blockchains start with a single block, called Block 0 or the Genesis Block. All other blocks are appended to this block, which acts as the ancestor to whom all other blocks may practically trace their genealogy because each block references the one before it. A cryptographic algorithm is used to link blocks, and a consensus architecture is employed to control and digitally sign transactions. New blocks are added to the network, carrying new information. Blockchain makes sure that everyone always sees the blockchain from the same perspective so that transparency is maintained. Since the information is open to all, anyone can confirm and validate it [5].

Blockchain enables immutable transactions, consensus, verification and validation. It is easily shared among several nodes without the requirement for any central authority because of its decentralised nature. The benefits of decentralised sharing of data include accountability, openness, validation and verification. Blockchain is more secure due to its immutability. Federated learning (FL) combined with blockchain offers a fresh method for decentralised data exchange. Consensus algorithms underpin blockchain's decentralised nature by ensuring that all users may trust the system without the need for a central authority. Multiple entities are involved in the management of blockchain. Such entities include nodes, validators, transactions, blocks, consensus, and service providers. Every entity has contributed to the decentralisation, security, functionality, and governance of the network. Blockchains are inherently decentralised networks comprised of multiple individuals who behave according to their available knowledge and incentives [6].

Consensus is attained when a single state is accepted by a large majority of the agents that comprise the network. The blockchain's security and integrity are upheld via a consensus algorithm. A consensus algorithm is a technique that enables all nodes in a distributed network to agree upon a single reality interpretation [7]. Consensus methods ensure uniform data viewing for all blockchain network participants and transaction validation in a secure, untrusted environment. If an algorithm meets both of the following two requirements, that is, agreement and termination, it is said to reach a consensus. In agreement, all non-faulty nodes choose the same output value. In termination, each non-faulty node selects an output value at some time [8]. Various consensus algorithms are Proof of Work, Proof of Stake, Proof of History, Proof of Activity, proof of authority, Proof of Burn, and Proof of Capacity. The choice of algorithm can have a significant impact on the network's scalability, security, and performance [9].

11.1.1 Blockchain Is Divided into Four Parts-Public, Private, Hybrid and Consortium Which Are Explained Below

Public Blockchain: It is also called the permissionless blockchain. Anyone can join and conduct operations without needing any permission from a central authority. Its characteristics include decentralisation, immutability, and transparency. Its disadvantages are scalability issues, more energy consumption and vulnerability to attacks. For example, Bitcoin and Ethereum.

Private Blockchain: It is also called permissioned blockchain. Work is conducted within a closed network. When developing the private blockchain application, the relevant authorities and chain developers selected the entities for this network. There are some limitations on who can take part in validations and transactions. Its characteristics are high performance due to fewer nodes, high throughput, scalability can be increased, speed is fast, etc. Its limitation is centralised, limited network effects, interoperability issues, etc. [10]. For example, Hyperledger.

Hybrid Blockchain: This technology blends public and private blockchain with permission-based and permissionless systems. It is designed to remove all the disadvantages of both blockchains. Members have the authority to determine who is allowed to use the blockchain and whose transactions are visible to the public. It is highly flexible while preserving integrity, security, and transparency. Its advantages are high transaction speed, reduced transaction cost, flexible infrastructure, etc. For example, Energy Web Foundation, Dragonchain, and R3.

Consortium Blockchain: Also known as federated blockchain. It is a type of private blockchain where the consensus process is controlled by a pre-selected group of nodes or organisations [11]. Consortium blockchains are operated and maintained by several groups or entities. Consortium blockchains are useful for use cases where several reliable and well-known parties need to work together on a similar platform because they strike a compromise between control and decentralisation [12].

Data cannot be altered once it is written on a blockchain due to immutability. Due to the decentralised structure of the blockchain network, a breach on a single node does not compromise the system. By facilitating standardised, secure, and efficient data transfer between diverse healthcare systems and providers, blockchain solves scalability and interoperability issues. The decentralised structure of blockchain eliminates single points of failure [3]. Blockchain can be used in healthcare to provide security and privacy to healthcare data and provide transparency to the healthcare.

11.1.2 Advantages of Blockchain in Healthcare

- **Collaboration and Information Exchange:** Blockchain can increase interoperability by reducing the standard exchange of health data between multiple systems and organisations. Authorised healthcare provider keeps and access the full patient record, guaranteeing that all pertinent data is available for improved care coordination.
- **Better Privacy and Security for Data:** Improved data privacy and security in the healthcare industry is essential for safeguarding private information, building patient trust, averting data breaches, and guaranteeing legal compliance. It not only safeguards patient data but also improves the general efficacy, efficiency, and reputation of healthcare providers, opening the door to better patient care and innovative medical practices since blockchain uses immutability to secure data.
- **Openness and Trackability:** Due to its decentralised nature, data is available to all nodes. So, it provides openness and trackability [13]. Every transaction is documented on the blockchain, which maintains a public, auditable record of all data access and modification operations. Blockchain technology ensures transparency and traceability throughout the pharmaceutical and medical supply lifecycle, from manufacturing to patient delivery.
- **Effectiveness and Economic Downsizing:** Smart contracts can be used to automate administrative tasks, including billing, processing claims, and managing patient records, which can save on overhead and administrative workloads. Blockchain can improve the efficiency of healthcare operations by streamlining a number of processes and reducing the number of delays.
- **Improved Studies and Clinical Investigations:** For regulatory compliance and public trust, it is essential to guarantee the integrity of clinical trial data. Trial data can be safely recorded and verified using blockchain. Blockchain technology may accelerate the processes of recruiting and obtaining consent from clinical trial participants, ensuring efficiency and transparency.
- **Preventing Fraud:** Blockchain technology can help in the identification and verification of patients and providers, decreasing the possibility of fraud in healthcare billing and services. The risk of counterfeit drugs can be decreased by tracking pharmaceuticals on the blockchain and verifying their validity.
- **Empowerment of Patients:** Patients own and manage their health data, deciding when and to whom it can be accessed. Increased patient involvement and engagement in their care can result from having more control over their health data.
- **Management of Population Health and Public Health:** Blockchain technology has the ability to instantly compile health data from several sources, offering insightful information for public health response and monitoring. While preserving individual privacy, blockchain technology can make it easier to use anonymised health data for research and population health studies.

- **Security of Medical Equipment:** Blockchain technology can be used to safely record and transfer medical device data, guaranteeing its integrity and security. Blockchain technology can be used to track and guarantee that medical devices are adhering to regulations.

11.1.3 Privacy and Security Concerns in Healthcare

Security and privacy are essential in the healthcare sector to protect sensitive information, build trust, and stop fraud and identity theft. Healthcare providers collect and keep vast amounts of sensitive data, including financial records, medical histories, and personally identifiable information. Safeguarding these data is necessary to ensure the confidentiality of patients. The reputations of healthcare providers are determined by their policies on the safeguarding of patient data. Breaches can severely damage relationships and trust in the workplace. In the healthcare sector, blockchain technology enhances data security, privacy, interoperability, and sharing. The privacy, security, and accessibility of patient information are all compromised when numerous medical systems are employed; apart from these, privacy and security have some disadvantages, like data leakage, complexity, possible delays in data access, and difficulties in finding a balance between security and use. The potential consequences of a security issue include availability, data integrity, and privacy [14].

Healthcare systems are still vulnerable to sophisticated cyberattacks like ransomware, advanced persistent threats and insider threats, even with strong security measures in place. Vulnerabilities that are not yet patched or newly identified can still be exploited. Ransomware attacks have the ability to encrypt important data, halting medical systems and affecting patient care. There is no assurance that access will be fully restored or that the data hasn't been altered even after attackers demand substantial ransoms to be paid in exchange for allowing access to be restored [15]. A replay attack is used to repeat the same request multiple times and pretend like a new request. It can be prevented by using nonce, timestamp, token expiration, encryption, etc.

11.2 PROBLEMS IN HEALTHCARE

11.2.1 Data Sharing Between Hospitals

Inadequate implementation of access control measures could lead to unauthorised access to private patient information. Employees or insiders who have been given access to examine data may misuse it, such as accessing the data modify the data, whether intentionally or accidentally, for their benefit. Due to the frequent compartmentalisation of healthcare data across multiple systems and organisations, information sharing is difficult and secure. Inconsistent data formats and standards may make it more difficult to integrate and exchange secure data [16]. When patient data is shared between hospitals for better diagnosis, a patient's data is encrypted using the receiver's public key and decrypted using the receiver's private key. After that, digital signature is used to identify users and protect information in digital messages. The sender signs the data using his private key, and the receiver authenticates using the sender's public key [17]. In this way, data is protected from unauthorised access and changes.

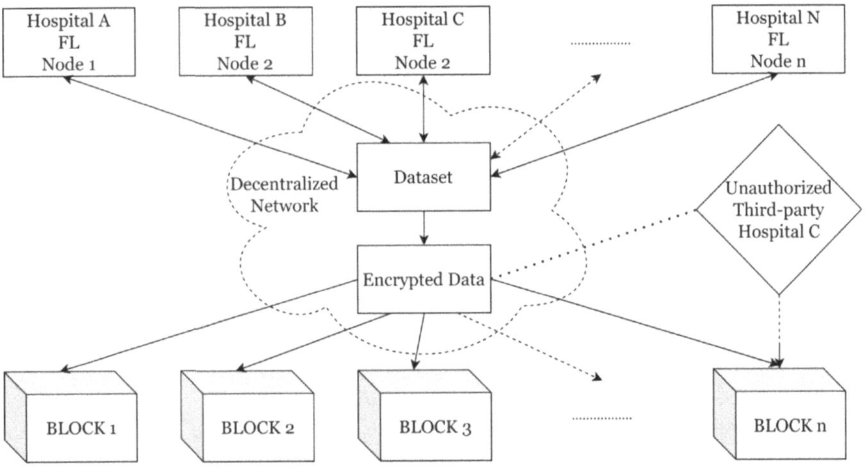

FIGURE 11.1 Data sharing in healthcare.

The above problem can be addressed as follows:
When data is shared from Hospital A to Hospital B, a malicious third party, such as Hospital C, is an unauthorised entity. It can attack the data, compromising system security, as shown in Figure 11.1. The following are the possible attacks:

- **Man-in-the-Middle Attack:** Hospital C puts itself in the way of Hospital A and Hospital B in order to intercept and maybe manipulate the data that is being transferred.
- **Replay Attack:** Hospital C listens in on the conversations that take place between Hospital A and Hospital B. In an attempt to reinsert data or start unsanctioned data requests, it attempts to resend intercepted transactions.
- **Identify Masquerade:** By faking signatures or changing its IP address, Hospital C tries to pass for Hospital A or Hospital B.

To prevent these attacks, several mechanisms can be proposed:

- **End-to-End Encryption:** The public key of the recipient is used to encrypt data when it is transferred between hospitals. This guarantees that Hospital C will not be able to decode or alter data without matching private keys, even in the event of a man-in-the-middle attack.
- **Hash and Nonce Verification:** A nonce value is included in each block to guarantee that the hash of the block corresponds to a specific pattern. As the blockchain checks the nonce and hash values, replay attacks are identified. A reused nonce value indicates a replay attempt.
- **IP Verification and Digital Signature:** If Hospital C poses as either Hospital A or Hospital B, the digital signature verification will not succeed. IP address verifications were also added as an additional layer of security to better foil identity masquerade attempts [2].

11.2.2 Share Sensitive Data

It is challenging for physicians and other healthcare professionals to communicate sensitive patient data because of the privacy-related characteristics of medical information. Healthcare models based on machine learning and deep learning are still not widely used due to a lack of realistic and up-to-date medical data.

The above problem can be addressed as follows:

Blockchain technology and FL allow for distributed learning between several physician collaborators while maintaining patient privacy. When a combination of FL and blockchain is used, it provides an extra layer of security. FL is also called collaborative learning. It is a machine learning technique that trains a global model across multiple decentralised servers holding local data samples without exchanging data [18]. Initially, weight is assigned by the global model to the local model. It trains the local models using a local dataset of that healthcare and sends the weights of all local models to the global model. All local model weights are encrypted and sent to the aggregator to aggregate weights using Secure Multiparty Computation (SMPC) and train the global model, as shown in Figure 11. 2.

SMPC, also known as MPC (Multiparty Computation), allows several parties to work together to compute a function using confidential data, allowing for precise "Blackbox" calculations that protect the privacy of individual inputs. SMPC, also known as privacy-preserving computation, safe computation, or MPC, is a transformational cryptography technique. Without ever disclosing their inputs to one another, it allows several people to compute a function using private inputs and display a public output.

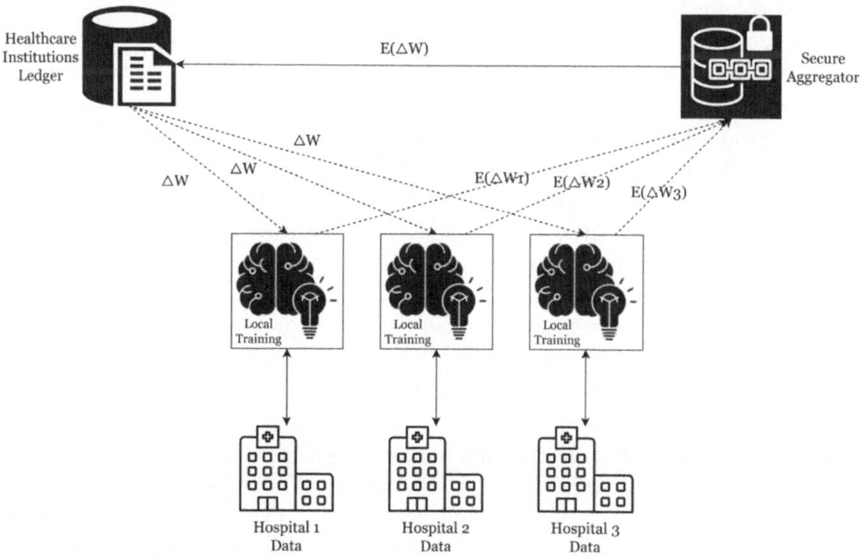

FIGURE 11.2 Healthcare system architecture.

Encryption and secret sharing are employed to guarantee confidentiality and privacy during the processing phase. Because it allows users to safeguard their data while still using shared systems, SMPC is crucial [19]. Without exchanging data, it allows numerous actors to construct a single, resilient machine-learning model, and it updates the weights of all local models. Again, the local model will be trained, and the global model will be trained without sharing the raw data. So, the privacy of healthcare data is preserved.

Therefore, it is possible to solve important challenges like access to heterogeneous data, data security, privacy and access rights. Suppose weights are directly sent to the global model from all local models without encryption. In that case, attacks that are possible using reverse engineering can obtain sensitive data from only locally generated updates, which is a problem [20]. So, to avoid this problem, the weight is encrypted and then sent to the global model. Many encryption algorithms (like homomorphic algorithms) can be used to encrypt the weight. Data encrypted with homomorphic encryption can be computed without the requirement first to decrypt it. Throughout the aggregation process, this feature is essential for preserving data privacy [21]. In this way, the global model is updated without sharing raw data. SMPC aims to train on heterogeneous datasets [22].

11.2.3 Healthcare Fraud

Healthcare fraud is an organised criminal activity in which false claims are filed with the help of peers, providers, doctors and beneficiaries. A comprehensive analysis of Medicare data has exposed a large number of fraudulent physicians. They use ambiguous diagnosis designations to support the priciest medications and therapies. Insurance companies are among the most vulnerable entities impacted by these unethical actions [23].

As a result, insurance companies raised their premiums and daily increases in healthcare costs. In the healthcare system, fraud and abuse can take many different forms. Some of the most common forms of provider fraud include service charges are never paid, Making false claims for identical service, Fabricating details on the services provided, charging for a more expensive or sophisticated service than was actually rendered, and Charging for a service that is covered even though it was not rendered. Hackers can utilise the medical information to locate the prescriptions while buying prescription medications online, which they can then use to commit fraud or identity theft [24]. In order to force victims to provide healthcare data, they would prefer to keep it secret, and hackers may also employ extortion and blackmail.

The above problem can be addressed as follows:

FL fraud enables FL-enabled policies through adaptive training and testing, which trains the model to detect fraud tendencies. The model can adjust to novel and unforeseen fraud patterns since it is self-learning. The fraud pattern was divided into two categories by the study: known patterns and unknown patterns. Patterns that have already been recognised and recorded are referred to as known patterns. They are comprehensible and frequently applied in forecasting and decision-making.

These trends can be utilised to guide expectations for the future because they have been seen in historical data. Patterns that have not yet been recognised or recorded are known as unknown patterns. To be useful, they need to be discovered and analysed because they are not yet understood. Machine learning and sophisticated data mining techniques are frequently used to find these patterns.

Initially, a pattern is given to every node. Before being executed by the global node, the applications transfer their workloads to the locally accessible fog nodes for additional processing (such as testing and training). The suggested approach first allocated a certain pattern to every task and then transferred them to the corresponding fog nodes that were accessible. If a new workload follows the same previous pattern, then it is not considered fraud. If it does not follow the same pattern, i.e., an unknown pattern, then send workload to another node. If any node matches the pattern, then it is not fraud. If the pattern does not match, that is an unknown pattern, then inform all nodes that this pattern is considered fraud. In this way, static fraud is detected.

The smart contract rule is used by the current proof of work and proof of stake consensus systems to conduct data fraud and validation amongst nodes in a blockchain-enabled network. The blockchain features machine learning-enabled random-forest fraud detection integrated to tackle dynamic fraud in the network. Dynamic and runtime unknown attacks can happen [25].

11.2.4 Single Factor Authentication

Additional crucial components of security include authentication and access control. Many healthcare systems continue to utilise password-only access and other single-factor authentication methods despite their susceptibility to theft and guesswork. Bad password practices that compromise security include using the same password over and over again and changing it infrequently [26].

The above problem can be addressed as follows:

To avoid this problem, blockchain uses FL, Internet of Things (IoT), cloud computing, etc. The single-factor authentication problem can be solved by implementing multifactor authentication (MFA). Due to MFA, the risk of unwanted access is considerably reduced. MFA, a multi-step account login process, requires users to provide more information than just a password that is used for OTP. Passwords protect digital assets, but they are not enough because proficient fraudsters look for passwords on a regular basis. Someone could be able to access multiple accounts that you may have used your password for if they figure out what it is. MFA offers an additional layer of security in the event that the password is compromised, preventing unauthorised individuals from accessing these accounts. Many healthcare facilities use MFA to verify user identities and provide authorised users with convenient access [27].

11.3 USE CASE AND APPLICATIONS

Application of privacy and security in blockchain

11.3.1 Electronic Health Record

The patient creates his ID and password. They submit their report on the portal, and available doctors will see the report. After consulting the doctor, the hospital will update the patient's profile. The doctor takes permission to get the patient's private key to access the patient's record, and the patient will provide his private key [1]. Only then will the doctor access the patient's data. In this way, the people who know the private key will access the data. So, privacy and security will be maintained.

11.3.2 Clinical Trails and Research

An immutable ledger created by blockchain technology can capture all clinical trial data, guaranteeing data integrity and regulatory compliance. In addition, it has the ability to manage patient consent efficiently and transparently [28]. Due to immutability, data privacy and security are maintained.

11.3.3 Medical Billing and Claim Processing

Through smart contracts, blockchain can automate the processing of invoicing and claims, lowering errors and administrative expenses while avoiding fraudulent claims [25]. The content or condition written in a smart contract will only execute in a sequence. Due to smart contracts, privacy and security are maintained.

11.3.4 Prescription Management

Blockchain technology can record every prescription transaction, ensuring that medications are authentic and given as prescribed. Consequently, there is a reduced probability of medication errors and prescription fraud [29] due to its immutability nature.

11.3.5 Challenges

The delicate nature of medical data, the complex nature of healthcare systems, and the constantly changing environment of cyber threats provide several obstacles to privacy and security in the industry. Some of the challenges are:

11.3.6 Cyberattacks and Data Breach

Ransomware: Attacks using ransomware, in which cybercriminals encrypt data and demand money in order to decrypt it, are especially frequent targets of healthcare organisations. Phishing: Phishing attempts deceive workers into exposing login details or downloading harmful software, which results in leaks of confidential information. Advanced persistent threats advanced persistent threats: long-lasting, highly skilled attacks are designed to steal confidential data slowly.

11.3.7 Problems with Interoperability

Interoperability is the ability of different blockchain networks and systems to communicate, share data and work together as a single entity. It is essential to allow various blockchain platforms to interact with and benefit from one another's features, which raises the general effectiveness, scalability, and usefulness of blockchain technology. Many issues pertaining to patient participation, security, privacy, consent, and governance arise with the move towards patient-centred interoperability. It is difficult to ensure smooth, safe data sharing between various healthcare systems and providers while maintaining individual privacy. Risks and costs may result from the absence of established protocols and formats for data interchange [30].

11.3.8 Insider Dangers

Workers who have access to private information may abuse it or do harm with it. Data breaches can result from employees' accidental behaviour, such as handling data improperly or falling for phishing attacks.

11.3.9 A Complex Regulatory Framework

Internationally functioning medical organisations are subject to a variety of administrative regulations, which can be difficult to manage and contradicting.

11.3.10 Outdated Systems

Many healthcare institutions still use outdated technologies, which are more difficult to secure and more open to attack. It can be difficult to integrate the latest, secure technology with older systems, being technically complex and costly.

11.3.11 The Volume and Complexity of Data

Healthcare systems create huge quantities of data, which makes effective security and monitoring challenging. Both structured and unstructured data—such as lab results, pictures and medical records—are included in healthcare data, and they all need various safety measures.

11.4 CONCLUSIONS AND FUTURE

In this Chapter, the security and privacy measures provided to healthcare data are explained. At first, several mechanisms to prevent the attacks that are possible in sharing data between hospitals are defined. By using several mechanisms to prevent possible attacks, data are easily shared between hospitals. Next, we have described how to train the global model without sharing the raw data to protect the privacy and security of data. In this, the local model is trained using the local dataset, weights are encrypted, and then the weights are aggregated via SMPC using homomorphic encryption and sent to the global model. The global model will be trained with the help of these weights, and aggregated values will be sent to the local model. Various strategies are discussed to prevent healthcare fraud by known and unknown patterns. If the upcoming pattern matches the existing pattern, then fraud has not occurred; if not, fraud has occurred.

Furthermore, MFA is performed to prevent single-factor authentication problems by adding an OTP facility along with an ID and password. In this way, privacy and security are maintained in healthcare. This book chapter explores the possibilities of integrating blockchain, FL, and cyber-security technologies for healthcare data management and security. The FL mechanism can be used in disease prediction models, personalised medicine, and novel therapeutics. Patients can benefit from more precise and customised diagnostic tools without having to worry about their privacy being compromised by using FL to train models using local hospital data.

The resilient and flexible fraud detection systems can constantly learn from a larger variety of data sources without disclosing private information. By enabling real-time fraud, cyber-attack and unauthorised access detection and prevention, advanced models may strengthen the healthcare system's defences against ever-changing threats. Along with more conventional techniques like an ID and password, future authentication systems may use advanced biometrics like fingerprints, face recognition, and others.

REFERENCES

1. M. Guduri, C. Chakraborty, U. Maheswari, and M. Margala, "Blockchain-Based Federated Learning Technique for Privacy Preservation and Security of Smart Electronic Health Records," *IEEE Trans. Consum. Electron.*, vol. 70, no. 1, pp. 2608–2617, 2024, doi: 10.1109/TCE.2023.3315415.
2. S. H. Alsamhi et al., "Federated Learning Meets Blockchain in Decentralized Data Sharing: Healthcare Use Case," *IEEE Internet Things J.*, vol. 11, no. 11, pp. 19602–19615, 2024, doi: 10.1109/JIOT.2024.3367249.
3. N. El Madhoun and B. Hammi, "Blockchain Technology in the Healthcare Sector: Overview and Security Analysis," In *2024 IEEE 14th Annu. Comput. Commun. Work. Conf. CCWC 2024*, pp. 439–446, 2024, doi: 10.1109/CCWC60891.2024.10427731.
4. N. Moosavi, H. Taherdoost, N. Mohamed, M. Madanchian, Y. Farhaoui, and I. U. Khan, "Blockchain Technology, Structure, and Applications: A Survey," *Procedia Comput. Sci.*, vol. 237, no. 2023, pp. 645–658, 2024, doi: 10.1016/j.procs.2024.05.150.
5. J. Golosova and A. Romanovs, "Overview of the Blockchain Technology Cases," *Proceeedings of 59th International* Scientific *Conference on Information Technology* and *Management Science* of *Riga Technical University* ITMS 2018, pp. 1–6, 2018, doi: 10.1109/ITMS.2018.8552978.
6. A. A. Siyal, A. Z. Junejo, M. Zawish, K. Ahmed, A. Khalil, and G. Soursou, "Applications of Blockchain Technology in Medicine and Healthcare: Challenges and Future Perspectives," *Cryptography*, vol. 3, no. 1, pp. 1–16, 2019, doi: 10.3390/cryptography3010003.
7. M. Xie, J. Liu, S. Chen, and M. Lin, "A Survey on Blockchain Consensus Mechanism: Research Overview, Current Advances and Future Directions," *Int. J. Intell. Comput. Cybern.*, vol. 16, no. 2, pp. 314–340, 2023, doi: 10.1108/IJICC-05-2022-0126.
8. R. K. Dewang, M. P. Yadav, S. Awasthi, O. Raj, A. Mewada, and K. L. Bawankule, "Data Secure Application: An Application That Allows Developers to Store User Data Securely Using Blockchain and IPFS," *Multimed. Tool Appl.*, vol. 83, no. 15, pp. 45491–45517, 2024.

9. V. Zhebka, S. Zhebka, T. Bazhan, P. Skladannyi, and V. Sokolov, "Methodology for Choosing a Consensus Algorithm for Blockchain Technology," In *CEUR Workshop Proc.*, vol. 3665, pp. 106–113, 2024.
10. Z. Liang, L. Baixiang, Z. Ruyi, J. Binxin, and L. Yijiang, "Overview of Blockchain Technology," *Jisuanji Gongcheng/Computer Eng.*, vol. 45, no. 5, pp. 1–12, 2019, doi: 10.19678/j.issn.1000–3428.0053554.
11. P. P. Ray, D. Dash, K. Salah, and N. Kumar, "Blockchain for IoT-Based Healthcare: Background, Consensus, Platforms, and Use Cases," *IEEE Syst. J.*, vol. 15, no. 1, pp. 85–94, 2021, doi: 10.1109/JSYST.2020.2963840.
12. K. K. Vaigandla, M. Siluveru, M. Kesoju, and R. Karne, "Review on Blockchain Technology: Architecture, Characteristics, Benefits, Algorithms, Challenges and Applications," *Mesopotamian J. CyberSecurity*, vol. 2023, pp. 73–84, 2023, doi: 10.58496/MJCS/2023/012.
13. Z. Wenhua, F. Qamar, T. A. N. Abdali, R. Hassan, S. T. A. Jafri, and Q. N. Nguyen, "Blockchain Technology: Security Issues, Healthcare Applications, Challenges and Future Trends," *Electronics*, vol. 12, no. 3, 2023, doi: 10.3390/electronics12030546.
14. M. A. Ansari and D. K. Singh, "Human Detection Techniques for Real Time Surveillance: A Comprehensive Survey," *Multimed. Tools Appl.*, vol. 80, pp. 8759–8808, 2021, doi: 10.1007/s11042-020-10103-4.
15. N. Spence, N. Bhardwaj, and D. P. P. Iii, "Ransomware in Healthcare Facilities: A Harbinger of the Future?," *Perspectives in Health Information Management*, no. November, 2018, [Online]. Available: https://www.proquest.com/docview/2111721098/415DFBB303EB44B8PQ/1?accountid=31923.
16. H. J. Tange and R. P. H. M. Smeets, "Information Exchange between Physicians and Nurses," *Comput. Methods Programs Biomed.*, vol. 43, no. 3–4, pp. 261–267, 1994, doi: 10.1016/0169–2607(94)90078-7.
17. S. M. Matyas, "Digital Signatures-An Overview," vol. 3, pp. 87–94, 1979.
18. M. Khan, F. G. Glavin, and M. Nickles, "Federated Learning as a Privacy Solution—An Overview," *Procedia Comput. Sci.*, vol. 217, pp. 316–325, 2022, doi: 10.1016/j.procs.2022.12.227.
19. C. Zhao et al., "Secure Multiparty Computation: Theory, practice and applications," *Information Science.* (Ny)., vol. 476, pp. 357–372, 2019, doi: 10.1016/j.ins.2018.10.024.
20. Y. Li, Y. Zhou, A. Jolfaei, D. Yu, G. Xu, and X. Zheng, "Privacy-Preserving Federated Learning Framework Based on Chained Secure Multiparty Computing," *IEEE Internet Things J.*, vol. 8, no. 8, pp. 6178–6186, 2021, doi: 10.1109/JIOT.2020.3022911.
21. F. Loukil, C. Ghedira-Guegan, K. Boukadi, and A. N. Benharkat, "Privacy-Preserving IoT Data Aggregation Based on Blockchain and Homomorphic Encryption," *Sensors*, vol. 21, no. 7, pp. 1–23, 2021, doi: 10.3390/s21072452.
22. Z. A. El Houda, A. S. Hafid, L. Khoukhi, and B. Brik, "When Collaborative Federated Learning Meets Blockchain to Preserve Privacy in Healthcare," *IEEE Trans. Netw. Sci. Eng.*, vol. 10, no. 5, pp. 2455–2465, 2023, doi: 10.1109/TNSE.2022.3211192.
23. W. El-Samad, M. Atieh, and M. Adda, "Transforming Health Insurance Claims Adjudication with Blockchain-Based Solutions," *Procedia Comput. Sci.*, vol. 224, pp. 147–154, 2023, doi: 10.1016/j.procs.2023.09.022.
24. M. Pareek, "Comparative Assessment of External Third Party Administrators and In-house Teams of Insurers in Health Insurance Claims Settlement Process in Bangalore, India," *J. Health Manag.*, vol. 26, no. 2, pp. 394–406, 2024, doi: 10.1177/09720634241229555.
25. A. Lakhan et al., "Federated-Learning Based Privacy Preservation and Fraud-Enabled Blockchain IoMT System for Healthcare," *IEEE Journal of Biomedical and Health Informatics*, vol. 27, no. 2, pp. 664–672, 2023, doi: 10.1109/JBHI.2022.3165945.

26. F. Ghaffari, E. Bertin, J. Hatin, and N. Crespi, "Authentication and Access Control Based on Distributed Ledger Technology: A Survey," In 2020 *2nd Conf. Blockchain Res. Appl. Innov. Networks Serv. BRAINS* 2020, pp. 79–86, 2020, doi: 10.1109/BRAINS49436.2020.9223297.
27. E. Ahmad and A. Alghamdi, "A Verification System for Multifactor Authentication for E-Healthcare Architectures," *Arab J. Sci. Publ.*, vol. 2663, p. 5798, 2021.
28. I. A. Omar, R. Jayaraman, K. Salah, I. Yaqoob, and S. Ellahham, "Applications of Blockchain Technology in Clinical Trials: Review and Open Challenges," *Arab. J. Sci. Eng.*, vol. 46, no. 4, pp. 3001–3015, 2021, doi: 10.1007/s13369-020-04989-3.
29. P. Li, S. D. Nelson, B. A. Malin, and Y. Chen, "DMMS: A Decentralised Blockchain Ledger for the Management of Medication Histories," *Blockchain Healthc. Today*, vol. 2, p. 38, 2019, doi: 10.30953/bhty.v2.38.
30. W. J. Gordon and C. Catalini, "Blockchain Technology for Healthcare: Facilitating the Transition to Patient-Driven Interoperability," *Computational and Structural Biotechnology Journal*, vol. 16, pp. 224–230, 2018, doi: 10.1016/j.csbj.2018.06.003.

12 Implementing Blockchain in the Industrial Internet of Things (IIOT)-Enabled Metal Casting

Divya Mobarsa, Amit Sata, Minal Shukla, and Yuri Nikolaichik

12.1 INTRODUCTION

Metal casting is possibly one of the earliest industrial processes. Its history extends back to 3,000 BCE. It eventually became indispensable in producing intricate parts that needed strength and strength, particularly for sectors like energy, aircraft, and automobiles. High-pressure die casting and lost-wax casting need to be updated to satisfy the efficiency and accuracy demands of the current world. Notwithstanding these advancements, there are still some extremely complex issues with operational efficacy, transparency, and data protection [1]. The Industrial Internet of Things (IIoT) greatly impacts managing these metal casting processes because it offers real-time monitoring. However, the system is susceptible to data manipulation and erroneous records from the manufacturing unit because of all this interconnectedness. However, the decentralization of IIoT systems may lead to deterioration in data integrity, reducing the dependability of supply chain activities. This chapter aims to provide an improved framework that should guarantee the authenticity and integrity of data transmission in an IIoT-enabled metal casting supply chain to address these issues. Blockchain technology, which was first developed for the Bitcoin business, is a good substitute that has been considered [2]. Blockchain provides the decentralized ledger system the immutability and security to solve all IIoT data handling issues. Smart contracts improve operations' efficiency, transparency, and traceability by automating the supply chain. By radically altering how the industry operates, incorporating blockchain technology into IIoT-enabled metal casting has the potential to improve production quality overall, lower costs, and strengthen the supply chain.

12.2 PRIOR WORK

The IIoT represents a digital transformation where traditional industrial systems are integrated with innovative digital technologies and enable rapid information exchange and decision-making by associating devices, equipment, and sensors in a

DOI: 10.1201/9781003545620-16

network. This relationship improves efficiency and productivity and enables better monitoring, control, and optimization of industrial activities. Some of the salient components of IIoT include sensor networks, clouds of computing infrastructure, and communication protocols. Industry 4.0 aims to optimise and sustain production processes through the Internet of Things (IoT), additive manufacturing, and robotics. This paper discusses the possibility of utilising blockchain technology and additive manufacturing to overcome the current challenges and make suggestions for future integration into the supply chain [1].

This chapter presents a blockchain-based IIoT platform to address security, trust, and connectivity problems in the IIoT industry. It broadcasts maintenance services and manufacturing equipment information using smart application scenarios [2]. This research presents a cooperative architecture for generating a database for the squeeze-casting process using blockchain technology to ensure data integrity, security, and on-demand access control [3]. It was found that the need to survey such escalated data generation in the industrial sector is necessary to facilitate informed decision-making and operational oversight. It combines IoT with machine learning to obtain environmental data from IoT sensors, applying big data techniques. The hybrid prediction model uses random forest classification to find anomalies and, thus, problems within the manufacturing system [4].

The combination of blockchain with IoT allows companies, such as utilities, engineering and construction, manufacturing, oil and gas, etc., to reimagine their industry structure flexibly. This study introduces a novel utilisation of blockchain technology. Blockchain technology and the IoT allow industries, including manufacturing, utilities, oil and gas, engineering, and construction, to restructure their organisational structures flexibly [5]. A monitoring and traceability system, controlled by a digital twin, is suggested for the general parts of a continuous casting machine. This system allows for predictive maintenance and the prediction of the remaining lifetime in situations with high temperatures and oil pollution [6].

This work focused on zero defect manufacturing detection approach and created a lightweight deep residual networks-based blockchain-enabled quality control platform. The system can function in real-time, resulting in a reduction in inference time by 2.2. A two-stage training procedure is devised to simplify the process, where the lightweight network is initially trained using feature maps obtained from a more intricate network. The approach was verified in two real-world scenarios, showcasing enhancements in F1-score and accuracy [7].

The article explores the incorporation of industrial DevOps and Integrating Industry 4.0 technology into the supply chain management system for a railcar manufacturer. The system employs systems thinking and a domain-driven design methodology to identify a multi-hierarchical structure, ultimately creating a real-time cloud manufacturing service system based on blockchain technology [8]. Industry 4.0 expedites industrial processes for customised, intelligent, and environmentally friendly production.

Nevertheless, these plans have been postponed due to security obstacles. These issues might be resolved by blockchain technology. However, research on its application in production processes has just recently started. This research examines how blockchain technology may be used while maintaining cost-effectiveness, flexibility, and manufacturing capacity. In addition, it offers a strategic plan for future investigation in this area [9]. The Fourth Industrial Revolution revolutionises corporate operations by necessitating sophisticated and scalable production facilities. Artificial intelligence (AI) and multiagent systems (MAS) techniques can be employed. MAS is a modelling approach that captures the interconnections between various production units. It enables the possibility of performing realistic pretesting without causing any disturbances to the ongoing operations.

This chapter examines how MAS and blockchain technology might improve knowledge and comprehension in a competitive setting [10]. Due to the rising prevalence of modular buildings, construction managers are emphasising off-site production management more. IoT-enabled Building Information Modeling (BIM) platforms provide enhanced visibility and cooperation. However, they frequently encounter obstacles such as a singular point of failure and the need to guarantee the origin of modifications made by diverse sources. This project aims to develop a platform called blockchain-enabled IoT-Building Information Modeling (IoT-BIM) for modular building [10].

This platform will address the constraints of current systems and facilitate the implementation of real-life applications. Subsequent research should prioritise optimising and evaluating the platform in diverse situations [11]. Industrial Cyber-Physical Systems, IIoT, digital copies, large quantities of data, and cloud computing are some of the sectors 4.0 technologies that are revolutionising the metal forming industry. These technologies are expected to enhance supply chains, raw material acquisition, tool design, production processes, energy efficiency, costs, quality assurance, and customer service [12].

The Steel Chain Protocol Suite aims to improve the steel industry's supply chain by utilising blockchain technology to create a standardised and transparent framework. This method correlates physical steel coils/rods with cryptographic transaction hashes, guaranteeing consistent integrity from start to finish [13]. This chapter examines the difficulties and issues encountered with blockchain-based IoT applications, focusing on IoT as a decentralised data source management and sensor solution. The text delves into the intersection of blockchain technology and IoT, examining necessary preparations, challenges, and potential future uses. In addition, it delves into the possibilities offered by blockchain regarding the IoT [14]. The article presents a framework designed to integrate blockchain technology into AI-enhanced industrial processes, highlighting the particular requirements and potential benefits it provides.

The essay analyses the capacity of AI and blockchain to enhance the efficiency and speed of product design, collaboration, and supply chain security [15]. Combining IIoT with blockchain in Industry 4.0 yields advantages such as real-time monitoring, inventory optimisation, and compliance with sustainability standards. It alleviates delays, inaccuracies, and intermediaries while promoting environmental, social, and economic benefits. This review examines prospective research directions [16]. The paper presents a blockchain-based system for effectively administrating building materials and inventory, enhancing monitoring capabilities, and ensuring transparency. This method promotes cooperation, mitigates material supply delays, and diminishes legal conflicts. This system combines economic order quantity with blockchain smart contracts to enhance decision-making and decrease material expenses [17].

The study examines the worldwide increase of electronic supply chains and related dependability challenges, proposing establishing a blockchain-based certificate authority system for effective chip data management. This system would mitigate several hazards, including recycling, remarking, cloning, and overproduction [18]. The research examines the operational efficiency and security issues within the mining industry in Industry 4.0, focusing specifically on the application of Blockchain and digital watermarking technologies. It introduces an innovative approach for executing these solutions on microcontroller devices [19]. The growing number of people using electric appliances leads to an increasing environmental impact and increased resource consumption. Circularising the Electrical and Electronic Equipment value chain may significantly reduce waste and promote sustainable practices. This method enhances the ability to monitor and trace objects while reintegrating raw materials into the system [20]. Collaboration in cloud manufacturing is dependent on trust. This study presents a trust mechanism named "Blocktrust" that utilises blockchain technology to tackle this problem. The private Blocktrust peer-to-peer network offers rapid transaction speed and imposes limitations on membership [21]. This chapter presents InterMat, a framework for discovering and exchanging materials data that utilises blockchain technology and combines cloud platforms. It facilitates the complete process of exchanging material data, encompassing data production, management, finding, sharing, tracking, and appraisal.

InterMat tackles difficulties associated with exchanging Integration of materials data from many platforms [22]. The chapter analyses manufacturing Small and Medium-sized Enterprises (SMEs)' challenges in implementing IIoT technologies and explores potential Blockchain technology solutions. It aims to guide SMEs in developing long-term digitisation plans and strategies [23]. Security services are used by blockchain-based IoT applications to leverage blockchain technology. However, it requires significant energy and computational resources, posing obstacles to BIoT implementation. Energy efficiency, security, throughput, latency, fork issues, security, legal concerns, smart contracts, storage, and network broadcast techniques are just a few of the difficulties this chapter examines when integrating blockchain technology with the IoT.

In addition, it makes recommendations for future study paths to deal with these issues and create effective BIoT applications [24]. The chapter presents a system framework for small and medium-sized intelligent factories utilising blockchain technology. This framework connects the quality control module of electrochromic material with the product traceability module, resulting in enhanced production quality and logistics of the supply chain. The progress of the electrochromic field is especially dependent on this [25].

12.3 FRAMEWORK

The metal casting process comprises various sub-processes, such as the creation of wax patterns and assembly, ceramic covering manufacturing, fettling, cleaning and finishing, testing and quality evaluation, dewaxing, heating, melting, and pouring. Each sub-step within the overall process necessitates the use of certain equipment for its execution. The metal casting process primarily involves using various specialised equipment and units. These include wax preparation units responsible for preparing the wax used to create the patterns. Wax injection presses, semi-automatic and automatic, inject the wax into moulds to create patterns. Ceramic slurry preparation units, which consist of equipment such as slurry mixers, drums, and rainfall sanders, are used to prepare the ceramic slurry. Autoclave units are used for dewaxing, collecting, and recycling wax. Pre-heating furnaces are used to heat the moulds before casting. Melting units, including melting furnaces, cooling towers, crucibles, and ladles, are used to melt the metal.

Fettling units, which consist of knock-out machines, shot blasters, welding equipment, and cutters, are used to finish the castings. Computer numerical control machines are used for in-house die-making and finishing processes. Lastly, quality inspection units are equipped with spectrograph equipment, universal testing machines, impact testing equipment, non-destructive testing equipment, microscopes, and measurement devices to ensure the quality of the castings. These machines often produce a significant amount of data on process parameters, but they cannot connect to the internet for collecting and streaming this information. This limits their ability to be used for process monitoring and data analytics [5]. A specialized series of IoT gateways may be designed to collect data from most of the units. Figure 12.1 illustrates the different materials and significant equipment used in the metal casting process.

Data analytics and process monitoring can benefit from acquiring machine data. Alternatively, the principles of blockchain may be employed to stream data to distributed servers instead of relying on the conventional practice of streaming data to a centralised server system, as commonly seen in current IIoT systems. Figure 12.2 illustrates the comprehensive approach incorporating IIoT and blockchain into the current metal casting foundry setup.

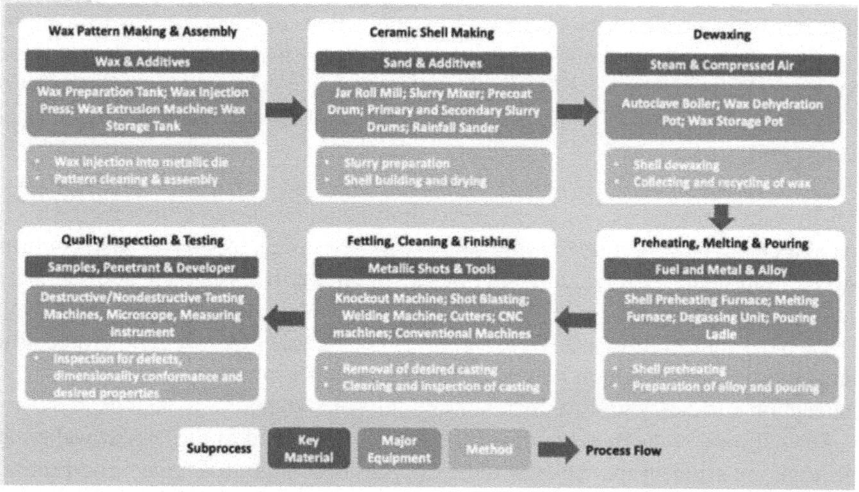

FIGURE 12.1 Existing metal casting process.

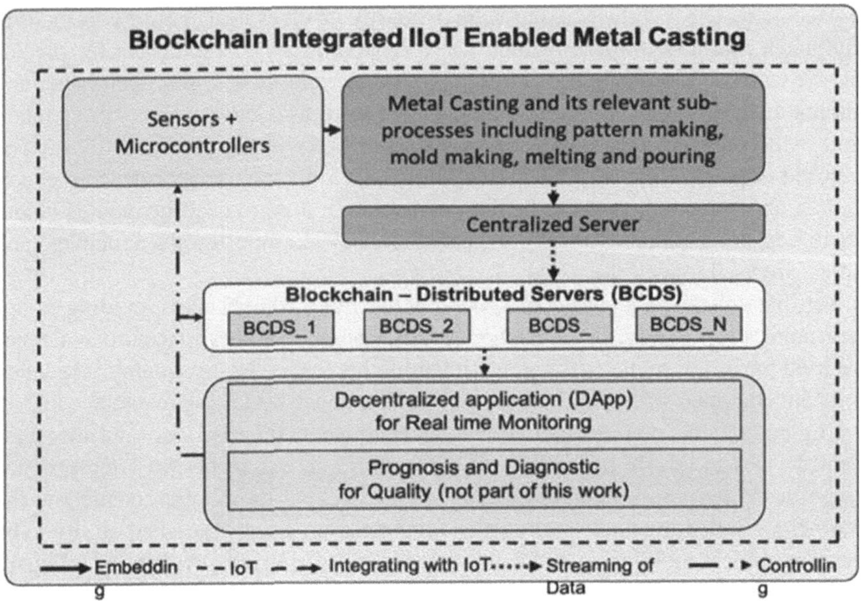

FIGURE 12.2 Integration of IIoT and blockchain with existing metal casting process.

12.4 IMPLEMENTATION

Earlier said, blockchain is an unchangeable, decentralised digital ledger system that has garnered significant attention for its capacity to keep a high degree of protection and anonymity. It operates without the need for any specific central authorities or repository. The use of suitable cryptographic hashes often safeguards transactions utilising the principles of blockchain. A hash function, which is a mathematical process, transforms an input numerical value into a condensed output numerical value. Any input length is processed by the hash function, which reliably yields outputs of a predetermined length. These outputs are then digitally signed and authenticated using public and private key pairs [8].

The transactional information is regarded as a block, a collection of transactions. It is then documented as a series of events that cannot be altered or modified without the agreement of the community responsible for recording, modifying, and confirming the transaction. Every individual block containing transactional information is linked to the preceding chain of transactions, ultimately becoming the blockchain. These blocks undergo validation using diverse consensus processes and protocols to authenticate the blockchain network. Blockchain has evolved significantly since its conception and can now store currency and information [6].

Furthermore, a distinct significance has been attributed to the digital wallet, an electronic instrument that enables individuals to conduct transactions. The primary features that emphasise the entire deployment of blockchain in the IIoT-enabled investment casting process are interoperability (the capacity of the IIoT system to interact with the blockchain system), transparency, traceability, and accountability [11].

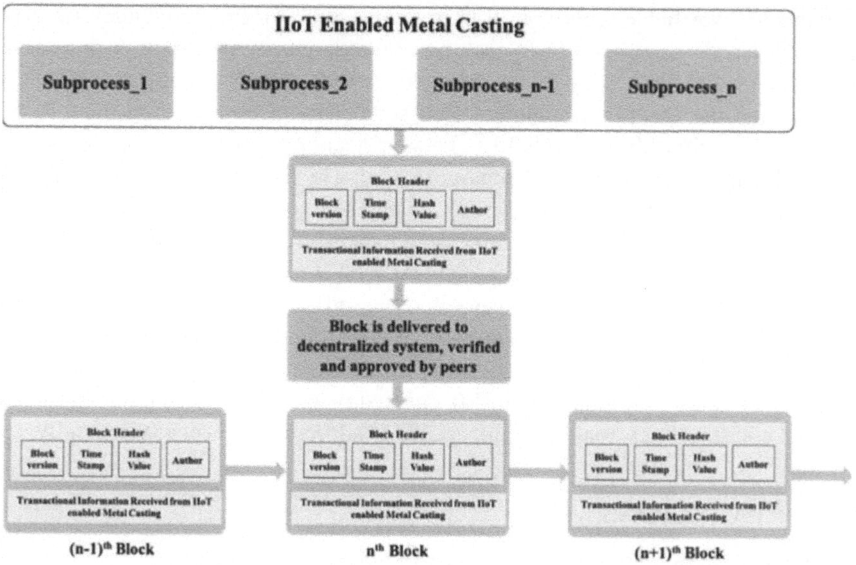

FIGURE 12.3 Blockchain implementation for IIoT-enabled metal casting.

Figure 12.3 illustrates the fundamental stages that demonstrate the progress of each block for every transaction that occurs in the IIoT-enabled metal casting process and its integration into Blockchain. Initially, sensors gather data at several points in the metal casting process, including temperature, pressure, humidity, and other pertinent factors. Next, the data is processed and structured to be inserted into the blockchain. The prepared data is securely delivered across the network, usually employing protocols appropriate for IIoT contexts.

Data integrity and authenticity are verified to ascertain that the acquired information originates from credible sources and has not been altered throughout its transmission. Subsequently, intelligent contracts are implemented according to pre-established criteria, automating specific elements of the metal casting procedure, such as quality assessments or scheduling. The validated data is organised into blocks, including an exact time and a secure, encrypted code. To ensure the block's integrity and stop alterations, it is encrypted. Consensus procedures based on Proof of Stake are employed to verify the authenticity of newly uploaded transactions on the blockchain.

Once a consensus is reached, a new block is added to the blockchain. The term "Blockchain" denotes a sequence of interlinked blocks with other blocks being connected to the preceding one. The blockchain is spread over several decentralised network nodes. Each system node maintains an exact copy of the blockchain, ensuring redundancy and fault tolerance. Cryptographic keys are employed to govern access to the blockchain. Permissions determine the specific rights that individuals have to view, write, or execute smart contracts on the blockchain.

At last, it is now feasible to do compliance audits on data stored on the blockchain.

Analytics may provide significant insights for reducing procedures and improving quality. Once data are added to the blockchain, it becomes immutable, guaranteeing a reliable and verifiable record of all transactions and process parameters. As mentioned earlier, data obtained from each piece of equipment are first sent to a server in the cloud and then sent to the blockchain to improve privacy and security. Broadcasting the records to a cloud-based server before broadcasting them to the blockchain is primarily motivated by the significant cost consequences associated with deploying a smart contract. Each transaction, which involves data gathered from each piece of equipment at second intervals and sent to a blockchain server, is processed directly without incurring excessive expenses associated with cloud-based servers. The cost per transaction is around 0.03 ETH (as shown in Figure 12.3). If a metal casting foundry does not utilise a cloud-based server at the intermediate level, the cost of this transaction will be quite high. The foundry must obtain 115 process parameters every second while producing an average of 40 tons of castings each month. Integrating blockchain with IIoT-enabled metal casting will have significant financial consequences, making it imperative to address this difficulty promptly.

12.5 DISCUSSIONS

The metal casting sector is being transformed by blockchain technology, which improves data integrity, security, traceability, transparency, supply chain optimisation, and quality assurance. It offers a distributed, unchangeable record for safe data transfer, removing the necessity for middlemen and trustworthy third parties. Blockchain technology promotes transparency in business activities, improves inventory management, and provides an immutable record for the metal casting process, reducing lead times and waste while maintaining quality control measures and compliance documentation. In blockchain technology, fraud and counterfeiting are prevented since they generate unique digital fingerprints for manufactured products, which assures the realness of products in industries such as automotive and aerospace. This further allows secure data interchange, increases cooperation, and protects intellectual property.

Blockchain systems, including IIoT, ensure smooth connectivity, reduce the cost of operations, and promote decentralised decision-making for localised decisions based on current facts. Since its inclusion in the IIoT framework, blockchain technology has altered metal casting significantly: it provides safety and traceability and leads to efficiency. However, there is concern about integrating it with present systems, scalability, and energy consumption. High-volume industrial operations might need private or consortium blockchains for better scalability. This decentralised, transparent, and irreversible platform deals with the problems that most traditional industries often face.

12.6 FUTURE SCOPE

IIoT for metal casting is investigating blockchain technology, which may find use in decentralised IIoT, regulatory compliance, economic impact, scalability, energy efficiency, privacy, and real-time monitoring, among other areas. Key operations in metal castings, such as supply chain coordination and production tracking, are being automated with smart contracts. Cyber security is essential for protecting critical manufacturing data, while data privacy and optimisation are also crucial. The main blockchain computing burden is being decreased by addressing scalability concerns and investigating Layer-2 solutions and consensus procedures. Green blockchain technology and energy-optimised consensus processes are two further areas of investigation that relate to sustainability and energy efficiency.

Distributed ledgers are used to provide immutable traceability and to expand real-time monitoring and traceability. Research is being done on regulatory compliance, standardisation, and interoperability throughout blockchain networks. Another area of research is the integration of AI and machine learning with blockchain, where AI models may be used to detect anomalies and optimise production parameters.

12.7 CONCLUSION

Metal casting incorporates blockchain and IIoT technologies to improve data security and integrity. Blockchain offers a decentralised ledger for industrial data traceability and tamper-proofing, and IIoT devices provide quality control and real-time monitoring. Agreements with stakeholders are automated using smart contracts, which guarantee payment upon fulfilment of output criteria. By combining IIoT with blockchain, manufacturers, suppliers, and consumers may build confidence, enhance product quality, and lessen data manipulation. Defects are decreased, production costs are decreased, demand forecasting accuracy is improved, and manufacturing delays are decreased. Smart contracts' real-time contract enforcement lowers disputes and boosts trust. By lowering operational costs and unplanned downtime, predictive maintenance increases uptime. This integration may speed up innovation and collaboration within the global metal-casting ecosystem. Accuracy, authenticity, and data reliability promote stronger commercial relationships. Blockchain and IIoT can boost production throughput, optimise resource utilisation, lower costs, improve sustainability, and increase efficiency. Adoption, however, can provide difficulties as it might affect standard practices and need adjustments from industry players.

REFERENCES

1. Ghimire, Turusha, Atharva Joshi, Samgeeth Sen, Chinmay Kapruan, Utkarsh Chadha, and Senthil Kumaran Selvaraj. "Blockchain in additive manufacturing processes: Recent trends & future possibilities." *Materials Today: Proceedings* 50 (2022): 2170–2180.
2. Singh, N., H. P. Singh, A. Mishra, A. Khare, M. Swarnkar, and S. K. Almas. "Blockchain cloud computing: Comparative study on DDoS, MITM and SQL injection attack." In *2024 IEEE International Conference on Big Data & Machine Learning (ICBDML)*, Bhopal, India, 2024, pp. 73–78. https://doi.org/10.1109/ICBDML60909.2024.10577412.
3. Deng, Jianxin, Gang Liu, and Xiangming Zeng. "Blockchain-based security access control system for sharing squeeze casting process database." *Integrating Materials and Manufacturing Innovation* 13.1 (2024): 92–104.
4. Shahbazi, Zeinab, and Yung-Cheol Byun. "Smart manufacturing real-time analysis based on blockchain and machine learning approaches." *Applied Sciences* 11.8 (2021): 3535.
5. Rane, Santosh B., and Yahya Abdul Majid Narvel. "Re-designing the business organisation using disruptive innovations based on blockchain-IoT integrated architecture for improving agility in future Industry 4.0." *Benchmarking: An International Journal* 28.5 (2021): 1883–1908.

6. Han, Zhoupeng, et al. "Digital twin-driven 3D visualisation monitoring and traceability system for general parts in the continuous casting machine." *Journal of Advanced Mechanical Design, Systems, and Manufacturing* 14.7 (2020): JAMDSM0100-JAMDSM0100.
7. Leontaris, Lampros, et al. "A blockchain-enabled deep residual architecture for accountable, in-situ quality control in industry 4.0 with minimal latency." *Computers in Industry* 149 (2023): 103919.
8. Matenga, Alice Elizabeth, and Khumbulani Mpofu. "Blockchain-based cloud manufacturing SCM system for collaborative enterprise manufacturing: A case study of transport manufacturing." *Applied Sciences* 12.17 (2022): 8664.
9. Song, Jinwoo, Jinbo Wang, and Young B. Moon. "Blockchain applications of manufacturing systems: a survey." ASME international mechanical engineering congress and exposition. Vol. 85567. American Society of Mechanical Engineers, 2021.
10. Kaswan, Kuldeep Singh, Jagjit Singh Dhatterwal, Sanjay Kumar, and Amit Pandey. "Industry 4.0 multiagent system-based knowledge representation through blockchain." In *Artificial Intelligence and Industry 4.0*, Aboul Ella Hassanien, Jyotir Moy Chatterjee, and Vishal Jain, eds. Academic Press, an imprint of Elsevier, Amsterdam, Netherlands, pp. 93–115, 2022.
11. Wu, Liupengfei, Weisheng Lu, Fan Xue, Xiao Li, Rui Zhao, and Maohong Tang. "Linking permissioned blockchain to the Internet of Things (IoT)-BIM platform for off-site production management in modular construction." *Computers in Industry* 135 (2022): 103573.
12. Liu, Heli, et al. "Industry 4.0 in metal forming industry towards automotive applications: A review." *International Journal of Automotive Manufacturing and Materials* (2022): 2–2.
13. Datta, Sreemana, and Parichay Bhattacharjee. "SteelChain—Blockchain-Based Transparent Supply Chain Framework for the Steel Industry." Advances in Thermal Engineering, Manufacturing, and Production Management: Select Proceedings of ICTEMA 2020. Springer Singapore, 2021.
14. Sharma, Arunima, and Ramesh Babu Battula. "Challenges and issues in blockchain-based IoT services." In *Blockchain-Based Internet of Things*, Debashis De, Siddhartha Bhattacharyya, and Joel J. P. C. Rodrigues, eds. Springer, pp. 47–80, 2022.
15. Patel, Dhruv, Chandan Kumar Sahu, and Rahul Rai. "Security in modern manufacturing systems: Integrating blockchain in artificial intelligence-assisted manufacturing." *International Journal of Production Research* 62.3 (2024): 1041–1071.
16. Soori, Mohsen, Fooad Karimi Ghaleh Jough, Roza Dastres, and Behrooz Arezoo. "Blockchains for industrial Internet of Things in sustainable supply chain management of industry 4.0, a review." *Sustainable Manufacturing and Service Economics* 3 (2024): 100026.
17. Basheer, Mohammad, Faris Elghaish, Tara Brooks, Farzad Pour Rahimian, and Chansik Park. "Blockchain-based decentralised material management system for construction projects." *Journal of Building Engineering* 82 (2024): 108263.
18. Xu, Xiaolin, Fahim Rahman, Bicky Shakya, Apostol Vassilev, Domenic Forte, and Mark Tehranipoo. "Electronics supply chain integrity enabled by blockchain." *ACM Transactions on Design Automation of Electronic Systems (TODAES)* 24.3 (2019): 1–25.
19. Evsutin, Oleg, and Yaroslav Meshcheryakov. "The use of the blockchain technology and digital watermarking to provide data authenticity on a mining enterprise." *Sensors* 20.12 (2020): 3443.
20. Fernandes, Leonardo, et al. "A Review on Adopting Blockchain and IoT Technologies for Fostering CE in the EEE Value Chain." (2023).

21. Vatankhah Barenji, Reza. "A blockchain technology-based trust system for cloud manufacturing." *Journal of Intelligent Manufacturing* 33.5 (2022): 1451–1465.
22. Wang, Changchang, et al. "InterMat: A Blockchain-Based Materials Data Discovery and Sharing Infrastructure." Processes 11.11 (2023): 3168.
23. Kumar, Ravinder, Rahul Sindhwani, and Punj Lata Singh. "IIoT implementation challenges: Analysis and mitigation by blockchain." *Journal of Global Operations and Strategic Sourcing* 15.3 (2022): 363–379.
24. Zafar, Saima, et al. "Integration of blockchain and Internet of Things: Challenges and solutions." *Annals of Telecommunications* 77.1 (2022): 13–32.
25. Wang, Qiyuan, Jingzhi Li, Muhong Zheng, Xie Ma, and Bin Wang. "Quality control and traceability framework of electrochromic materials based on block chain from the perspective of practical application." *Computational Intelligence and Neuroscience* 2022.1 (2022): 7376168.

13 The Role of Blockchain in Drug Traceability and Data Integrity

Nimmagadda Chandra Paul, M. L. S. N. S. Lakshmi, B. Gangadhara Rao, and Sumit Badotra

13.1 INTRODUCTION

Blockchain technology is a decentralized digital ledger that keeps an immutable, transparent, and secure record of transactions on many computers. Each block of the blockchain contains a list of transactions, and each transaction is connected with the one before it to form a chain. It ensures data are tamper-proof because once recorded, no amount of transactions can be altered without affecting all the subsequent blocks. Blockchain technology is used for data security, administrative efficiency, and interoperability to solve problems in the healthcare industry. It provides a transparent system and decentralization to improve drug traceability, guarantee the integrity of clinical trial data, and facilitate the management and exchange of electronic health records (EHRs). Other health sectors where blockchain technology can be applicable are shown in Figure 13.1.

In 2008, an unknown person began using Satoshi Nakamoto and presented blockchain technology as defined in the white paper "Bitcoin: A Peer-to-Peer Electronic Cash System." The first application of blockchain technology was the base for Bitcoin, a decentralized cryptocurrency allowing peer-to-peer trades without needing an intermediary. With this expansion, blockchain technology extended its focus from merely applying currencies to other fields, including supply chain management, finance, and healthcare. The emergence of platforms such as Ethereum introduced the concept of smart contracts, which are self-executing with the terms of an agreement written directly into code. This innovation has impacted others by automating and securing many healthcare operations, including insurance claims. Understanding blockchain's fundamental ideas and workings is crucial to fully appreciating its influence on the medical field. Blockchain maintains an immutable and transparent record of transactions across numerous computers. Every transaction is tracked in a "block," connected to every other block to form a "chain." So, the structure guarantees that every data is timestamped and verified and avoids tampering. In the next section, we shall discuss the objectives of blockchain technology in the healthcare sector.

 DOI: 10.1201/9781003545620-17

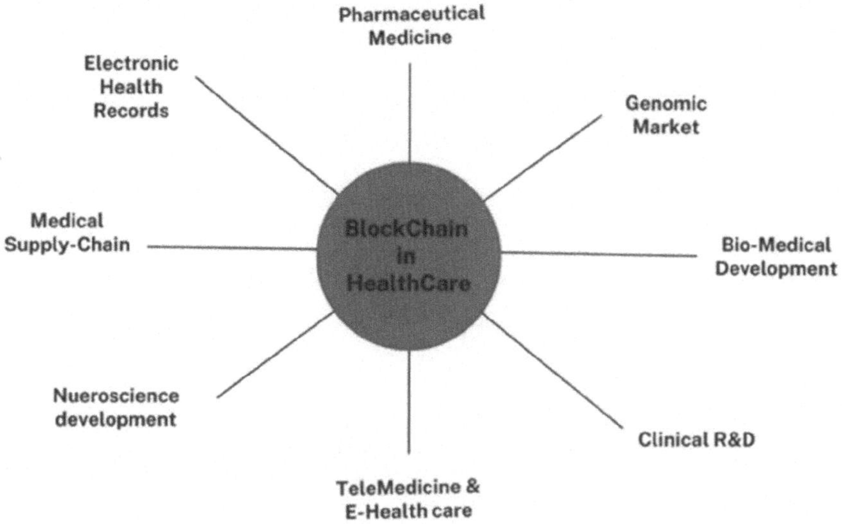

FIGURE 13.1 Blockchain technology in health care sectors.

13.2 OBJECTIVES

The objectives of blockchain technology in the healthcare sector are shown in Figure 13.2, briefly discussed in this section. Blockchain uses cryptographic algorithms to protect data, which makes it very hard for third parties and unauthorized parties to alter data. Blockchain can enhance the safety of medical data by shielding medical records from breaches, tampering, and unwanted access. Advanced cryptography algorithms offered by blockchain technology ensure the confidentiality and security of medical records. For example, Guardtime partners with the Estonian government in its effort to use blockchain to secure more than a million health records. Guardtime's KSI maintains a continuous log of all changes and accesses, which produces an immutable audit trail that minimizes data breach risk and may be able to detect malicious access attempts rapidly [1]. Another illustration is the blockchain-based Health Nexus platform by SimplyVital Health, which enables the safe sharing of patient data between healthcare providers.

The platform encrypts all patient data before it is stored on a decentralized network, and it also guarantees that only those who have permission can access it. We can reduce the danger by decentralization for centralized data storage, which also improves overall data security [2]. The Food Drug Administration (FDA) is also researching using blockchain technology to preserve clinical trial data. The FDA can ensure trial data's immutability and that data remains secure, accurate, and true to itself by keeping it on a blockchain. Changes are nearly impossible because every data entry is hashed and connected to the previous one. Such examples illustrate how blockchain technology can potentially bring much-needed security to health information in healthcare as it will prevent unauthorized access and further assure patient confidentiality and integrity [3].

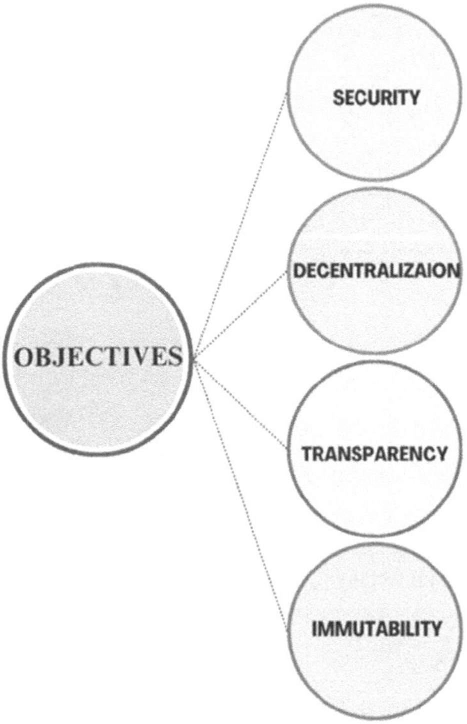

FIGURE 13.2 Applications of health care in the blockchain domain.

Fewer single points of failure and the distribution of control across multiple nodes have made systems more resilient, eliminating the necessity of a central authority.

Furthermore, blockchain technology allows for the centralized data collection maintained by all network users and, as a result of preventing any party from gaining control over the complete data collection, makes for decentralization. A medical chain, for example, would be a decentralized network of EHRs for exchange and storage using blockchain technology. This method grants patients full control of health information and allows access to medical professionals based on necessity. This decentralized method eliminates the possibility of data breaches and illegal access by avoiding the storage of health records in a single vulnerable location [4]. For example, IBM collaborated with the Centers for Disease Control and Prevention using blockchain to track public health data. Decentralization in gathering and sharing data improves the security and resilience of health information. Health information is ensured always to be available even if a few of the network nodes fail or are compromised [5]. In addition, the decentralized network that SimplyVital Health has developed, the Health Nexus, provides a safe exchange of patient data between them and other healthcare providers. Such a network reduces the risk of a data breach and makes sharing data much more efficient because it eliminates the need for one central repository [2]. Another process through which confidence between patients, providers, and stakeholders is promoted is transparency in health processes and transactions.

Blockchain technology creates an open, adaptable record of every action, ensuring that each transaction is viewable only to authorized parties on all activities. A good example of this in real-time is a blockchain-based project meant to enhance the pharmaceutical supply chain's transparency: MediLedger. MediLedger provides traceability and verifiability at every step in the production and distribution chain because every transaction is recorded on the blockchain. In this regard, transparency helps improve the trust value along the supply chain because it ensures that patients get authentic prescriptions and also aids in the destruction of counterfeits [6].

Another is the clinical trial blockchain application. The FDA and other organizations are researching blockchain to track and validate clinical trial data. One can use the blockchain to log the trial data; thereby, any alterations will be transparent and traceable, allowing the stakeholders to ensure the integrity and transparency of the data. Apart from the reduced chance of fraud or data tampering, this increases confidence in the result [3].

Furthermore, on its turn, MyClinic.com is one of the telemedicine networks based on blockchain technology. Blockchain will provide transparency in interactions between patients and health providers. Patients can ensure their data are handled properly and openly because they can view who accessed their medical records and the reasons for these interactions [7]. Once uploaded, ensure that the data are permanently and impenetrably recorded on the blockchain. This ensures that it cannot be removed or altered once the data are uploaded. A basic component of blockchain technology is immutability, thus enhancing the reliability and integrity of the data. For instance, the use of blockchain technology to ensure the correctness of data during clinical trials is an example currently being done. The FDA has been studying blockchain technology to create immutable records of clinical trial data. A timestamp is given to each entry on the blockchain, and then it is linked to the previous entry, making it almost impossible to alter undetected. It keeps tampering over data under control and provides confidence in research findings by ensuring that results from a clinical trial are true and reliable [6].

Another example is MedRec, an EHR system developed by MIT's blockchain. For a permanent health record, MedRec would ensure that data is submitted so it cannot be changed or removed. This immutability decreases the chances of medical errors and promotes care for patients as healthcare professionals are provided with accurate patient histories [4]. Furthermore, the blockchain is utilized by self-sovereign identification startup Evernym to build indestructible patient IDs that will ensure patients' authenticity and actual patient records. In the next section, we discuss various applications of blockchain in the healthcare industry.

13.3 APPLICATIONS OF BLOCKCHAIN IN HEALTHCARE

Blockchain technology has several promising applications in healthcare, as shown in Figure 13.2, and it aims to address challenges like security, interoperability, and data integrity. Some key applications are discussed briefly in this section. These applications highlight how blockchain can enhance transparency, security, and efficiency across various facets of healthcare, from patient care to supply chain management.

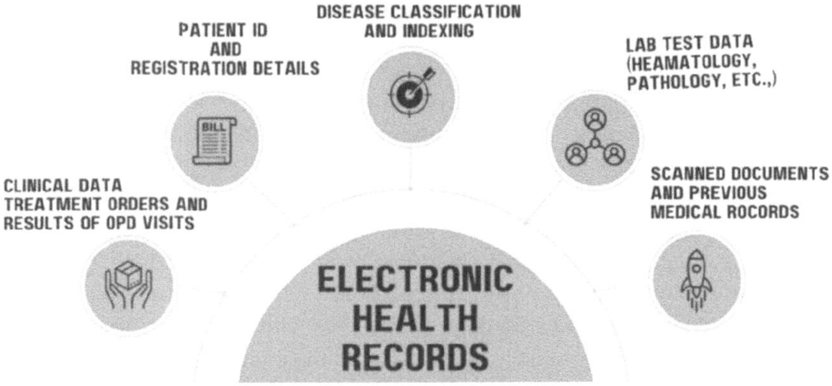

FIGURE 13.3 Maintaining electronic health records of patient data.

13.3.1 Electronic Health Records (EHRs)

The main purpose of EHRs is to securely maintain patient health records in a decentralized manner, as shown in Figure 13.3, which includes patient data, disease records, and lab test reports. This way, blockchain increases data accuracy and access and protects patients' privacy by making EHRs accessible and unchangeable. Blockchain technology introduces a new way of dealing with EHR issues by enhancing data security, interoperability, and patient control. One such project is MedRec, which researchers at MIT developed. MedRec has developed a decentralized network to store EHRs using blockchain technology.

In such an architecture, patients can self-manage their health information and provide permissions for access to other healthcare providers as needed. MedRec maintains records on a blockchain and offers an avenue through which patient data is safely encrypted, immutable, and accessible only by approved parties while addressing privacy concerns and lowering the danger of data breaches [4]. Another case is between the blockchain firm Guardtime and the UK National Health Service, NHS in London. Here, a blockchain system is intended to protect patients' data while ensuring that EHRs are accurate and unhackable. This technique logs every access to and modification of an EHR and creates an irreversible and transparent audit trail that enhances the security and trustworthiness of data [8].

Furthermore, the startup SolveCare has built a blockchain-based HER management and patient care coordination platform. Their system uses blockchain technology to create an open, secure way to track patient interactions and medical records. This increases the system's effectiveness by ensuring patients have control over their health information and may share it with many providers without hassle. This also ensures accurate data [9].

13.3.2 Supply Chain Management

Blockchain technology helps trace drugs from producers to consumers, thus reducing the possibility of spurious drugs entering the system. Because blockchain technology incorporates an immutable and transparent record of all transactions, it can considerably enhance the healthcare supply chain's administration. A prime example is the MediLedger, a blockchain-based medicine traceability tool. MEDILEGGER follows the whole pharmaceutical supply chain from its manufacturer to the end user. MEDILEGGER makes any point in a chain public and traceable since any transaction made is recorded on the blockchain. Blockchain creates transparency, which protects patient safety and builds trust in the pharmaceutical supply chain by helping to prevent counterfeit pharmaceuticals, eliminate fraud, and speed up recalls [6]. The health sector in the Food Trust blockchain project is an example of the relationship between Walmart and IBM. The traceability of medicines and food products improves when this network is used. Walmart can respond more promptly to its safety concerns and recall-related issues due to this technological advancement, which is adopting blockchain in real-time monitoring of the origin of drugs and foodstuffs. Walmart ensures customers have authentic and safe health products by increasing visibility within its supply chain [5].

A blockchain system has been developed for tracking and managing medical devices. The VeChain platform offers assurance that the safety and quality requirements in every step of the product life cycle are fulfilled based on real-time information regarding the location and status of the devices. This blockchain-based tracking system allows healthcare professionals to gain accurate and reliable information about their devices, preventing spurious or low-quality devices from entering circulation [10].

13.3.3 Health Data Interoperability

Health data interoperability aims to ease data flow around healthcare providers and systems. The blockchain allows decentralization control and standardized data formats, facilitating information flow in the health sector. It provides a decentralized and standardized platform for the exchange of health information, offering a powerful response to an enhancement in the interoperability of health data. One of the best-known is the Health Information Exchange network developed by blockchain firm SolveCare. This network uses blockchain technology to create decentralized, universal health-data ledgers. It enables seamless interoperability across multiple healthcare systems, reducing data silos and enhancing care coordination, as seen in Solve. Care ensures that patient information is always shared properly between different service providers, as this can be achieved through a single, immutable record accessible to other authorized organizations [9]. This is another example of using the MedRec platform and integrating the EHR systems within various hospitals. MedRec shares and manages the EHRs between different medical systems using blockchain technology. No matter where the information was generated, this integration ensures that the medical records of patients are up-to-date and available and allows for the standardization of health data. The blockchain-based approach reduces interoperability issues and simplifies care transfer among healthcare providers. The basic objective of the Mayo Clinic and BlockMedx blockchain venture is to enhance the interoperability of prescription medication data by using a blockchain-based technology that allows the tracking of prescription data safely and transparently across different pharmacies and healthcare providers. The system increases data consistency and accessibility, reduces the risk of medication errors, and increases patient safety by guaranteeing consistent and easy prescription information [7].

13.3.4 Patient-Centric Care

The notion of patient-centred care is such that it puts individuals in control and ownership of their health data. Blockchain empowers patients to enhance their autonomy and involvement in management through their health data since they can offer selected providers access to that data. Blockchain technology enables patients to seize greater control and access to their medical records, a major contribution to patient-centred treatment. One good example is the blockchain technology applied in MyClinic. It is a patient-centred health record system that allows patients to manage their EHRs personally through MyClinic and share them with other health practitioners whenever necessary. This way, patients are empowered to take charge of their health information, share it easily, and determine their treatment course [7]. Another such is the blockchain-based Patientory. With Patientory, patients can now own and manage their health data on a secure blockchain network. Patients still have discretion over who accesses their data and can easily share it with clinicians, researchers, or other organizations for direct care or paramount research projects. This empowerment makes patients more involved in care decisions and experiences greater engagement [11]. Health Nexus is a blockchain-based platform developed by SimplyVital Health company that focuses on patient-centric care coordination. With the use of Health Nexus, patients are enabled to track their real-time care interactions and control who accesses their health information. Two of the advantages that lead to better outcomes of more efficient and personalized treatment from this approach include more patient engagement in their health management and better-coordinated care among different providers [2].

13.3.5 Medical Device Management

The objective of medical device management is to track the use and maintenance of medical devices. Blockchain can generate an open record of how the devices have been used, maintained, and performed, thereby enabling regulations for the safety and management of efficient devices. Blockchain technology offers an indecipherable, unalterable record of every device's life cycle, including its manufacturing phase, throughout its various periods of use, and up to its end-of-life. This will enhance medical device management. Such noteworthy examples include VeChain's blockchain technology, which tracks medical devices in real-time, as shown in Figure 13.4.

Every device is precisely registered on a blockchain using VeChain's technology, thus giving rise to an exhaustive history regarding production, distribution, and use. This traceability aids in confirming regulatory compliance, stopping counterfeiting, and confirming the legitimacy of medical products [10]. Modum uses blockchain technology. Modum captures the state of medical devices in the storage process and transports them with the Internet of Things sensors and blockchain technology. The

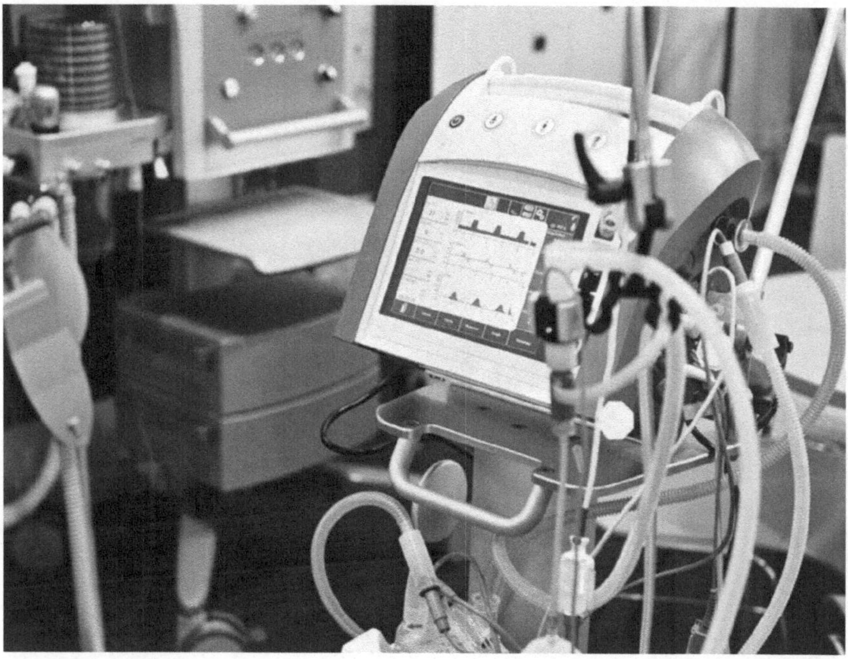

FIGURE 13.4 Blockchain technology can monitor the medical devices used for medical checkups of patients.

parameters to be captured by the sensors include temperature and humidity, with the blockchain holding the data. This approach provides safe and efficient medical device storage and transportation under ideal conditions [12]. Patientory is a startup that has developed an application based on blockchain that controls and follows the medical equipment in healthcare areas. Patientory technology ensures on-time medical device maintenance and servicing because of real-time tracking support. The blockchain ledger improves operational effectiveness and compliance because of an immutable and transparent record of device upkeep, repairs, and usage [11].

13.3.6 Telemedicine

Protection of patient information and telemedicine exchanges. Blockchain technology can protect patient data and create trust in remote patient care by ascertaining the safety and integrity of telemedicine sessions. The application of blockchain technology has an advantage in telemedicine because it offers a transparent and safe structure for managing patient data and communication. Of course, one worth mentioning is Healthereum, in which blockchain is utilized for secure virtual consultations between medical practitioners and patients and monitoring of participation from patients. Ethereum makes use of blockchain technology to create a decentralized ledger for all encounters made with telemedicine. This would ensure that patient information will be kept private and that access is limited to those allowed. Since everything recorded is recorded on camera and cannot be erased, this transparency

FIGURE 13.5 Doctors assisting the patients through telemedicine sessions.

fosters trust among patients and doctors [13]. It is a blockchain-powered platform where MyClinic uses blockchain technology for virtual consultations and ensures that the message between patients and medical professionals is safe and secure. MyClinic stores the data of every patient with consultation history in a blockchain, thus making all information safe, unchangeable, and accessible only to those who have provided their express consent. This enhances the total patient experience in telemedicine, as shown in Figure 13.5, and ensures the safety of patient data [7].

The blockchain-based healthcare platform SolveCare, which also supports modern telemedicine systems, uses blockchain to monitor and manage patient schedules, consultations, and medical data. This way, patients can monitor their personal health information during treatment using telehealth, and all interactions within the telehealth system are stored securely. SolveCare enhances patient activation using blockchain technology to strengthen telemedicine services, making them more secure and transparent [14]. Thus, these use cases reveal how blockchain technology may enhance security transparency and efficiency in various fields while dealing with many issues the healthcare sector faces.

13.4 CHALLENGES AND LIMITATIONS OF BLOCKCHAIN IMPLEMENTATION IN HEALTHCARE

Implementing blockchain in healthcare has significant challenges and limitations despite its potential to improve security, interoperability, and data integrity. Here are the main hurdles:

13.4.1 Technical Challenges

i. **Scalability and Performance Issues:** Serious scalability issues in blockchain technology stem from its consensus mechanisms and the requirement for each block to be copied to every node in the network. This can hinder blockchain networks from meeting the speed and performance demands of the healthcare industry. For instance, processing real-time health data on the Ethereum blockchain may be challenging due to its limitations in transaction throughput and latency despite its widespread use [15].
ii. **Integration with Existing Systems:** Integrating blockchain with current healthcare systems and EHR platforms presents significant challenges and costs. Many healthcare institutions rely on outdated systems that are unlikely to be compatible with blockchain technology. For example, linking blockchain to an existing EHR system would require significant transformation to ensure data integrity and interoperability, which can be technically complex and expensive [4].
iii. **Data Storage Limitations:** The amount of data that each block in the blockchain can hold is fixed, making it impractical to store large health records directly on the blockchain. Instead, hashes or references to off-chain data are stored on the blockchain. However, this approach may not adequately address the need for comprehensive data storage in the healthcare sector and can complicate the management and retrieval of holistic health records [16].

13.4.2 Regulatory and Compliance Obstacles

i. **Navigating Health Care Regulations:** Health care is one of the industries with the highest regulatory standards, besides regulatory standards ensuring strict privacy and security regulations. One may have to go through a rather cumbersome regulatory environment to apply blockchain in healthcare, the US legislation being considered under the Health Insurance Portability and Accountability Act (HIPAA). These blockchain solutions should be designed with the regulations and the special needs of managing healthcare data in mind [17].
ii. **Ensuring Data Privacy and Security:** Blockchain can ensure increased security by relying on cryptographic techniques, while problems exist in protecting user privacy. Sensitive personal information that should be protected can be found in healthcare data [8]. Questions arise about the right balance that needs to be struck regarding privacy and transparency in certain healthcare contexts because blockchain technology makes secrecy necessary.

iii. **Addressing Legal Implications of Blockchain Use:** The legal implications of using blockchain in medical areas are still under consideration. Questions like data ownership, the legality of an immutable record, and who is to be blamed if it is suspected that some data breaches have occurred must be addressed. Modifying or deleting patient health data that has been input error in a blockchain will be difficult, raising some legal dilemmas [18].

13.4.3 Stakeholder Resistance and Adoption Barriers

i. **Cultural Resistance to Change:** The adoption of advanced healthcare technologies is often culturally opposed. Since blockchain technology is relatively new and somewhat difficult to understand, a healthcare organization [19] and experts may be less likely to adopt it. The resistance will be dealt with justifiable proof of the advantages and adequate support of adopting blockchain-based systems.
ii. **Costs and Investments in New Technologies:** There are tremendous development, integration, and maintenance costs associated with adopting blockchain technology. The investment cost of blockchain technology could be prohibitive for healthcare companies if the benefits of its adoption show up too late [20]. Training costs, system updates, and compliance could significantly impede serious adoption.
iii. **Need for Training and Education:** To use blockchain effectively in healthcare, one needs to have a deep knowledge of blockchain technology and its applications. For this purpose, adequate training and preparation are also necessary for health providers to work with blockchain technology effectively [21]. The training programs should cover technical and practical areas of blockchain applications within healthcare services to ensure effective implementation.

13.5 FUTURE TRENDS IN BLOCKCHAIN HEALTHCARE TECHNOLOGY

Blockchain technology promises significant advancements in the healthcare sector in the coming years, with developments that could fundamentally transform various areas, as illustrated in Figure 13.6.

Integrating artificial intelligence [22,23] can enhance data analytics, thereby improving decision-making. Developing interoperability standards [16] will facilitate seamless communication between different health systems. Two privacy-preserving technologies (zero-knowledge proofs and secure multi-party computation) can potentially enhance data security. Furthermore, decentralized clinical trials [15] are expected to reduce costs and increase data transparency. It will ensure individuals are in charge of their health records because patients will have the central aspect through blockchain patient-centric models, and the smart contracts will simplify routine administrative tasks, reducing cost instead; productivity will be high. It will

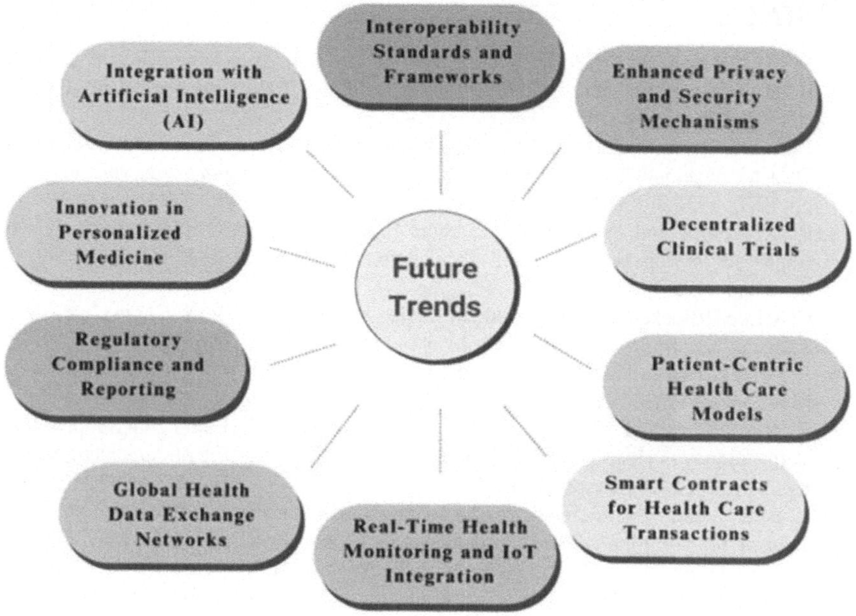

FIGURE 13.6 Future trends for healthcare in blockchain.

integrate with Internet of Things devices to make real-time health monitoring possible; international networks for health data exchange [7] will facilitate the sharing of health data beyond borders to respond better to the health needs of the public. Personalized medicine [24] will advance through safe genetic data exchange and blockchain's ability to ease regulatory compliance and reporting [25]. Patterns shown relate to and indicate that blockchain technology provides a promising platform for improvement in many aspects: patient-centred care, data security and protection, interoperability, and overall efficiencies in the health sector. Such innovations may facilitate greater outcomes in health.

13.6 CONCLUSION

Blockchain technology has the enormous potential to completely transform the healthcare industry by addressing its core issues, which include patient-centred care, data security, and interoperability. The decentralized, unchangeable structure offered by blockchain guarantees the confidentiality, transparency, and integrity of medical data and transactions. Practical implementations show how blockchain enhances security and productivity in the medical field. Guardtime's safe eHealth solution, Patientory's secure patient data management, and MediLedger's prescription tracking are among these. Scalability issues and regulatory compliance are some remaining issues, but these can be resolved with increased innovation and legislation supporting them.

REFERENCES

1. M. M. Hassan, J. P. McDaniel, and M. Zulkernine, "Blockchain-Based EHR for Improved Data Management in Health Care Systems," *IEEE Access*, vol. 7, pp. 104290–104305, 2019.
2. J. Zhang, H. Yang, and Q. Zhao, "Blockchain-Based Privacy Preservation for Health Data Management," *IEEE Transactions on Biomedical Engineering*, vol. 66, no. 7, pp. 1920–1931, Jul. 2019.
3. R. K. Gupta and A. Jain, "Blockchain-Based Health Information Exchange for Improved Interoperability," *IEEE Transactions on Biomedical Engineering*, vol. 68, no. 4, pp. 1052–1060, Apr. 2021
4. C. S. Lee, "Blockchain-Based Medical Device Management: The VeChain Case," *IEEE Transactions on Biomedical Engineering*, vol. 68, no. 7, pp. 2096–2103, Jul. 2021.
5. J. D. O'Connell, "Modum: Integrating Blockchain with IoT for Medical Device Monitoring," *IEEE Access*, vol.8, pp. 132456–132465, 2020.
6. R. A. Williams and M. J. Klein, "IBM Food Trust Network: Enhancing Supply Chain Transparency in Healthcare," *IEEE Transactions on Engineering Management*, vol. 68, no. 2, pp. 1056–1063, Apr. 2021.
7. R. B. K. Shah, "Healthereum: A Blockchain-Based Platform for Telemedicine," *IEEE Transactions on Biomedical Engineering,* vol. 67, no. 5, pp. 1342–1351, May 2020.
8. A. K. Patel and M. B. Puri, "Patientory: Blockchain-Based Medical Device Tracking and Management," *IEEE Journal of Biomedical and Health Informatics*, vol. 24, no. 10, pp. 2887–2895, Oct. 2020
9. M. A. Ansari and D. K. Singh, "Human Detection Techniques for Real Time Surveillance: A Comprehensive Survey," *Multimedia Tools and Applications*, vol. 80, pp. 8759–8808, 2021. https://doi.org/10.1007/s11042-020-10103-4.
10. R. K. Dewang, M. P. Yadav, S. Awasthi, O. Raj, Arvind Mewada, and K. L. Bawankule, "Data Secure Application: An Application That Allows Developers to Store User Data Securely Using Blockchain and IPFS," *Multimedia Tools and Applications*, vol. 83, no. 15, pp. 45491–45517, 2024.
11. Xie Y, Zhang J, Wang H, Liu P, Liu S, Huo T, Duan Y, Dong Z, Lu L, Ye Z, "Applications of Blockchain in the Medical Field: Narrative Review," *Journal of Medical Internet Research*, vol. 23, no. 10, pp. e28613, 2021.
12. P. Shelke, N.P. Sable, S. Dedgaonkar, and R. Mirajkar, "Applications of Blockchain: A Healthcare Use Case," In *Nature-Inspired Methods for Smart Healthcare Systems and Medical Data*. Springer Nature, Switzerland, pp. 67–88, 2023.
13. Zhuang, Yu et al. "Applying blockchain technology for health information exchange and persistent monitoring for clinical trials." *AMIA Annual Symposium Proceedings*. Vol. 2018. American Medical Informatics Association, 2018.
14. Gaynor, Mark, Kathleen Gillespie, Allison Roe, Erica Crannage, and J. E. Tuttle-Newhall. "Blockchain Applications in the Pharmaceutical Industry." *Blockchain in Healthcare Today* vol. 7, 2024.
15. A. M. Stout, "The Challenges of Blockchain Technology in Healthcare," *IEEE Journal of Biomedical and Health Informatics*, vol. 24, no. 4, pp. 1140–1147, Apr. 2020.
16. C. S. Haynes, "Legal and Regulatory Issues in Implementing Blockchain in Healthcare," *IEEE Consumer Electronics Magazine*, vol. 9, no. 5, pp. 78–85, Sep. 2020.
17. R. P. Chiu and M. J. Malan, "Economic Implications of Blockchain Adoption in Healthcare," *IEEE Engineering Management Review*, vol. 47, no. 2, pp. 60–70, Jun. 2019.

18. P. Zhang, J. White, D. C. Schmidt, G. Lenz, and S. T. Rosenbloom, "FHIRChain: Applying Blockchain to Securely and Scalably Share Clinical Data," *Computational and Structural Biotechnology Journal*, vol. 16, pp. 267–278, 2018.
19. N. Goenkaa, A. K. Sharmab, S. Tiwaric, N. Singhd, V. Yadave, S. Prabhuf, and K. Chadaga, "A Regularized Volumetric ConvNet Based Alzheimer Detection Using T1-Weighted MRI Images," *Cogent Engineering*, vol. 11, no. 1, p. 2314872, 2024. https://doi.org/10.1080/23311916.2024.2314872.
20. M. A. Engelhardt, "Hitching Healthcare to the Chain: An Introduction to Blockchain Technology in the Healthcare Sector," *Technology Innovation Management Review* vol. 7, no. 10, pp. 22–34, 2017.
21. T. T. Kuo, H. E. Kim, and L. Ohno-Machado, "Blockchain Distributed Ledger Technologies for Biomedical and Health Care Applications," *Journal of the American Medical Informatics Association*, vol. 24, no. 6, pp. 1211–1220, 2017.
22. M. Chen, J. Yang, Zhou, Y. Hao, J. Zhang, and L. Hu, "Blockchain Is Used for Secure and Efficient Data Sharing in the Healthcare Sector," *IEEE Access*, vol. 7, pp. 147674–147687, 2019.
23. Q. Xia, E. B. Sifah, A. Smahi, S. Amofa, and X. Zhang, "BBDS: Blockchain-Based Data Sharing for Electronic Medical Records in Cloud Environments," *IEEE Access*, vol. 6, pp. 53924–53936, 2017.
24. K. Bajaj, D. K. Singh, and M. A. Ansari, "Autoencoders Based Deep Learner for Image Denoising," *Procedia Computer Science*, vol. 171, pp. 1535–1541, 2020.
25. S. Ahmad, I. Shakeel, S. Mehfuz, and J. Ahmad, "Deep Learning Models for Cloud, Edge, Fog, and IoT Computing Paradigms: Survey, Recent Advances, and Future Directions," *Computer Science Review*, vol. 49, p. 100568, 2023.

14 Boost Financial Efficiency and Security with Blockchain

K. Sankar, Balajee Maram, G. Uma Devi, Pravin R. Kshirsagar, and Shaik Saidhbi

14.1 INTRODUCTION

A chapter titled "Boost Financial Efficiency and Security with Blockchain" discusses in-depth how blockchain technology further changes financial operations with a huge amount of transparency, fraud-free, and data security. Discussions behind this cover how blockchain helps to drive the streamlined financial processing process because of its decentralized nature, which reduces costs and offers trust places like finance and supply chains [1].

14.1.1 Importance in Financial Sector

The use of blockchain technology can significantly impact the financial industry by reconsidering the manner of performing and/or registering financial operations. This has been underlined with a few key factors:

Improved Security: Embezzlement loss, cyber-attacks, and other unlawful intrusions are some of the challenges that typical systems of finance encounter. These risks are minimized due to the decentralized and cryptographically secure feature of blockchain. Each of the transactions ends up in a record, which makes several ledgers spread across the network. Thus, in case of a need for alteration, it is close to impossible without the consent of the network. This improves data integrity and the possibility of fraud reduction [2].

Transparency and Traceability: Blockchain, being an open, immutable record of all transactions, creates that type of transparency that allows parties within a network to access the same view of data. Thus, all transactions are very transparent and verifiable. For financial institutions, this would mean enhanced accountability and the ability to trace transactions to a far greater degree than at present—something very important for compliance and auditing purposes [3].

Efficiency and Cost Savings: Traditional financial transactions usually have many intermediaries, such as clearinghouses and correspondent banks,

DOI: 10.1201/9781003545620-18

which extend both the length and cost of a transaction. Blockchain eliminates these while making it possible to have peer-to-peer transactions. This reduction in the number of middlemen can very easily lower transaction fees, expedite settlement times, and make the overall process easier [4].

Innovation with Smart Contracts: Due to the fact that smart contracts are self-executing, with the agreement terms directly written into lines of code, complex financial agreements can be automated. Such automation will reduce manual processing and oversight, minimize human error, and ensure that contract terms are executed according to the agreement—thus improving efficiency and reliability [5].

Access and Inclusion: Blockchain can help deliver better financial inclusion since it can allow more people to obtain access to financial instruments. Actually, it is entirely possible to engage in financial activities using the blockchain without having a bank account; what is needed is a digital wallet [6].

Blockchain technology responds to most of the current financial sector's major problems and provides better safety, higher transparency, and increased efficiency. Increasingly, adoption is going to transform financial operations and services in a manner that seeks to encourage innovation and possibilities of new opportunities in the field.

14.2 BLOCKCHAIN TECHNOLOGY BASICS

Blockchain is basically a decentralized, distributed ledger system that records transactions across many computers. Such a system ensures that data integrity and security are guaranteed because no one can alter records unless there is a consensus in the network. Main features include transparency, immutability, and the use of cryptographic techniques to secure data [7].

14.2.1 How Blockchain Works

Blockchain refers to a digital, decentralized, and fully transparent database that produces a record of dealings in a network of computers. All of these transactions are validated and grouped into a "block" and linked to a "chain" of preceding transactions, making it a continuous and absolutely unchangeable process [8]. Every block details a timestamp, transaction data, and a cryptographic hash of the previous block, linking them together. This cryptographic linkage ensures that the alteration of a previous block would oblige the recalculation of all subsequent blocks, which is computationally infeasible without network consensus. The transactions are verified and validated by way of a consensus mechanism—either Proof of Work or Proof of Stake—which requires agreement from a number of participants on the network before a block is added to a chain. This, therefore, makes the blockchain resistant to tampering and fraudulent activities because of its integrity and reliability, which is backed by cryptographic security in a decentralized process. The working process is shown in Figure 14.1.

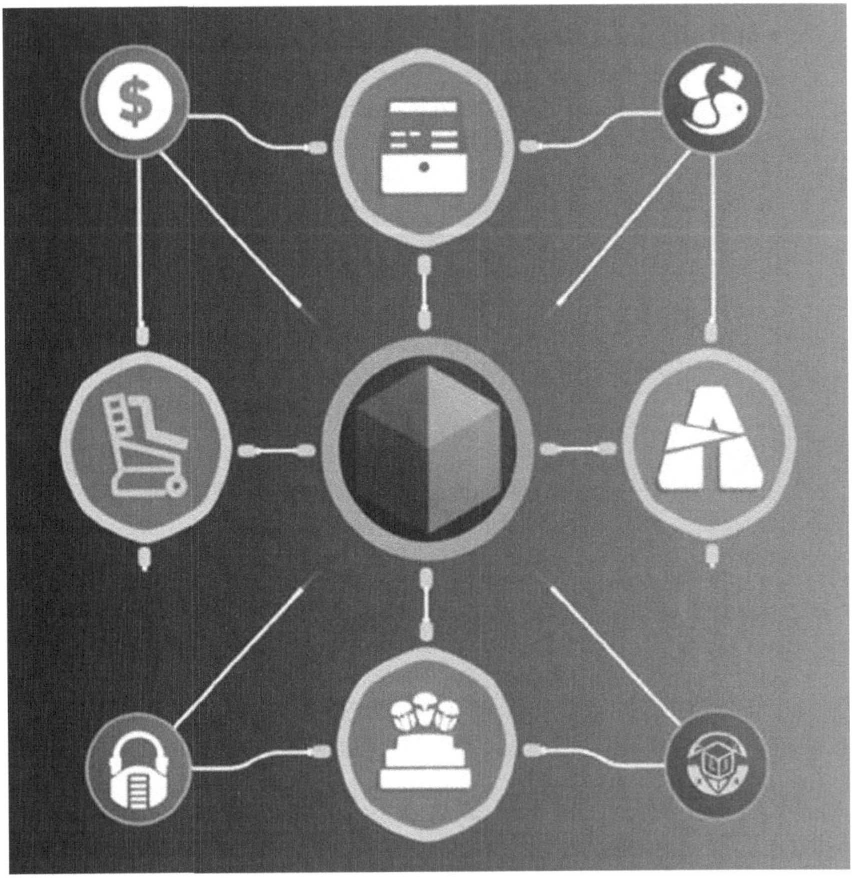

FIGURE 14.1 The working of blockchain technology.

14.3 BENEFITS THAT BLOCKCHAIN BRINGS TO FINANCIAL SERVICES

To begin with, blockchain technology benefits financial services because it makes them more transparent, secure, and efficient. It reduces intermediaries or even removes them because transactions become faster and cheaper. It also ensures that records are secured and tamper-proof, hence mitigating the risks of fraud. Automation processes by smart contracts also ensure the smooth flow of payment and lending, asset management, and all other operations [9].

14.3.1 Fraud Prevention with Blockchain Technology

The features that are inherently available in blockchain technology make fraud prevention much easier. The reason is related to the decentralized, immutable, transparent, and cryptographically secure nature accorded to the transactions. All of these features provide a secure and trustworthy environment, hence reducing fraudulent activities [10].

Decentralization: Traditional systems are weak and may lead to fraud because they have one point of control that could be susceptible to manipulation. Blockchain's decentralized nature distributes data across nodes in a network with no central point of authority over the network. Due to this decentralization, any single entity finds it painfully hard to manipulate the system—thereby reducing the possibility of fraud.

Immutability: A transaction, once recorded, cannot be altered or deleted from the blockchain. This property makes all transaction records permanent and tamper-proof. For someone to modify a transaction recorded previously, every subsequent block would have to be updated; this would be computationally infeasible because the majority of the network has to reach an agreement. This factor makes it completely impossible for fraudsters to tamper with transaction history or cover traces [11].

Transparency: This is where the blockchain becomes an open book, where every transaction happening within it can easily be viewed by the participants in the network. In simple words, all actions are open to easy auditing and verification, which makes it very hard for fraudulent activities to be passed unnoticed. This brings about a possible trace for fraudulent transactions, continuous monitoring, and real-time verification that would help nip it in the bud in financial systems.

Cryptographic Security: Blockchain creates advanced cryptographic techniques to secure data. All the transactions get encrypted and are then connected with the previous one through cryptographic hash functions. These linkages build a chain of transactions resistant to tampering and fraud. Besides, digital signatures using public and private key pairs ensure that only authorized users can initiate transactions and avoid unauthorized access and fraudulent activities [12].

Smart Contracts: Smart contracts are self-executing agreements with the terms directly written into lines of code (functions) on a blockchain, ensuring automatic execution without the need for intermediaries. They automatically enforce and execute a contract once criteria are met. As a result, automation decreases the necessity for an intervening party while massively decreasing the potential for human error and ensuring that transactions occur exactly as contracted, meaning there is less opportunity for fraud in the earnest execution of contracts [13].

Auditability: Transparency and immutability of a blockchain provide an auditable record of everything that happens on the chain. As such, it plays a significant role in the investigation and detection of fraud. An unbroken and auditable ledger of all transactions is opened up to auditors or regulators for rapid identification and prevention of fraud.

Consensus Mechanisms: These are used by blockchain networks as a verification system for transactions. These mechanisms needed a consensus of multiple entities in the network in the closing of moved transactions to their ledger. This process, which is validated collectively, ensures that no fraudulent transactions are accepted for validation, and only valid transactions go on to the blockchain [14].

Decreased Counterparty Risk: Blockchain lowers counterparty risk a lot as there may be complete trust on both sides of the business transaction, and all parties involved have identical information on the blockchain. This type of shared visibility and trust also reduces the risk of fraud in multi-party transactions such as trade finance or supply chain management [15].

Fraud Prevention with Blockchain Technology: Blockchain is built on a decentralized and immutable ledger that is transparent and cryptographically secure for users. These features dramatically enhance the security and integrity of digital transactions, making blockchain highly effective in fraud prevention.

Transparency and Traceability: Blockchain enables real-time tracking and tracing of products or transactions across supply chains. Its immutable ledger allows all involved parties to verify the authenticity and origin of goods, fostering accountability and eliminating fraudulent claims or activities. This builds trust across the entire chain, promoting transparency and integrity [16].

14.4 THE MERITS OF A PUBLIC LEDGER

The benefit of increased transparency, security, and trust among different applications in the financial realm offered by blockchain technology can be summarized in a public ledger. In other words, transparency allows everyone in the network to access the transaction history and verify transactions

It delivers accountability, which is how trust gets built. This is especially true in an environment where two opposing parties need to rely on each other, for example, in supply chain management and financial services [17].

The less vulnerable and most difficult to attack side of the decentralized public ledger makes it a more secure nuclear deterrent: users. All transactions are encrypted, linking the adjacent transaction chain with each other in such a way that it is nearly impossible to modify or change without the agreement of the network. This will reduce the possibility of tampering, fraud, and unauthorized access.

14.4.1 Immutable

Once data has been written to a public ledger, it cannot be erased, and it cannot be altered. It adds integrity and permanence to transaction records, and hence, it helps in making the history accurate and verifiable. This would enforce integrity over any past data because, this way, no single organization can change, through a backdoor, the history to commit fraud or cover their actions [3,18].

14.4.2 Trust and Decentralization

Public ledgers operate in a distributed system, which is a peer-to-peer system depending on the consensus of the participants, but there is no central authority or neither a third party facilitating the process. This quite decentralized approach is useful to foster trust among all participants because they all rely on consensus mechanisms to effect validation of the transactions. In the complete network, no single entity will have control, hence offering a simple method of detection and making the alteration unattainable in the design of the system, thereby making the system more democratic and secure [19].

14.4.3 Auditability and Traceability

Public ledgers are clear and cannot be altered, which presents a definite verifiable paper trail they offer. Each transaction is timestamped and then referred back to previous transactions to track the move of assets and to validate each transactional truthfulness to auditors and regulators. It is very helpful in compliance and forensic accounting.

14.4.4 Lower Costs and Higher Efficiency

Public ledgers can decrease the cost of transactions and be more effective since any third party is eliminated, and people are provided with P2P transactions. The techniques of financial transactions through intermediaries take their fees and time for processing the transactions. These are all made possible through the public ledger, hence reducing the associated costs and increasing the rate of the transaction processes [20].

14.4.5 Accessibility and Inclusivity

Public ledgers are available to any general Internet user, making financial inclusion common practice. It creates an avenue through which individuals and organizations, especially individuals in the unbanked and under-banked economies, can interface with the global economy. In this regard, accessibility may have a positive effect of encouraging innovation and growth by placing more power on individuals to participate in financial operations.

14.4.6 Improved Contract Execution

This enhances the efficiency and reliability of contract implementation in many contract-type applications.

Still, it is also important to note that public ledgers provide relatively high levels of benefits that stem from the levels of transparency, security, trust, immutability, auditability, efficiency of operations, accessibility, and capacity to execute contracts. This makes them a rather powerful means of increasing the reliability and effectiveness of financial and other related systems and applications.

14.4.7 Audit Trails in Blockchain Technology

Audit trails are essential in any information system that handles sensitive information or financial data, as they provide accountability, transparency, and integrity of the transactions. Blockchain technology offers very robust audit trail capabilities through the inbuilt features of immutability, transparency, and decentralization. This explains how blockchain enhances audit trails:

Immutability: A blockchain itself is immutable. The transactions recorded on the ledger cannot be deleted or changed.

Transparency: Since blockchain functions as an open book, every transaction taking place is visible to each participant within the network. The transparency of each transaction guarantees tracing and verification by each eye. Thus, it creates trust and accountability. For auditors, it actually offers real-time access to a ledger that contains a full and accurate history of transactions, greatly simplifying the auditing process and consequently reducing the risks of fraud or misreporting.

Decentralization: The blockchain is a decentralized approach to processing, forwarding, and storing data across a number of nodes in the network, with no single point of control or failure.

Timestamping: This is very useful in auditing, as it helps the auditor trace through the chronological order of events and verify the time of any transaction.

Efficiency and Automation: Through smart contracts, blockchain can also perform an element of the auditing process. Smart contracts are automated contracts where the conditions are hard-coded to execute themselves. It is an agreement executed through electronic platforms whereby pre-established requirements determine when the contract has been implemented and documented. Such automation reduces the need for manual verifications, reduces errors, and has indeed enhanced audit effectiveness.

Blockchain can be considered safe and sturdy for building and consolidating audit trails with the help of the current generation's technology. This technology is based on qualities such as unchangeability, openness, distribution, timestamping, secured through cryptography, capability of tracing, effectiveness, and a board with regulators that enhance the quality and credibility of audits. With these features, blockchain will be useful in establishing accountability, given its ability to constrain fraudulent activity in multiple sectors, notably within the financial sector and or any field that requires stringent auditing.

14.5 COMPARATIVE RESULTS AND DISCUSSION

A comparative analysis of various tools applied, representing some of the salient features, advantages, and disadvantages, is shown in Table 14.1.

14.6 CHALLENGES AND RISKS

Despite its merits, blockchain technology has challenges and risks. For one, scalability issues affect how fast transactions can be completed, which significantly increases cost. Another aspect is more related to regulation uncertainty since that may create

TABLE 14.1
Comparative Results

Criteria	Boost Financial Efficiency and Security with Blockchain	Enhancing Financial Transparency with Blockchain Technology	Blockchain for Finance: Enhancing Security and Reliability of Transactions	Utilizing Distributed Ledgers for a More Efficient and Secure Financial Performance
Relevance to trends	Highly relevant to financial markets and technology trends, addressing both efficiency and security.	Focuses on transparency, a key feature but narrower than efficiency and security.	Addresses the same efficiency and security but is more descriptive.	Relevant but "distributed ledger" is a more technical term, less known outside specialized circles.
Clarity and simplicity	Crystal clear; hence, it will be rather easily readable by a wide range of readers.	Simple, but in this variant, the accent is on just one aspect-transparency	Descriptive and slightly longer, although still relatively clear.	The presence of technical terms, such as "distributed ledger," makes it less readable for a general audience.
Focus	Balanced between the two core values of blockchain in financial efficiency and security.	Focuses solely on transparency, less balanced	Similar balance, but a bit more detailed between security and efficiency.	Strong focus on optimization, but in a very technical aspect-dependent way; for example, distributed ledger
Target audience	Target audience/ appropriate for technical and non-technical focused finance professionals	The audience would be more towards regulators and transparency-related audiences.	Likely has a similar target audience but with perhaps a little more in-depth insight.	Highly technical audiences comfortable with topics related to blockchain and financial systems
Potential impact	High impact because it resolves two major pains in finance: efficiency and security	Moderate; while transparency is a big issue, it is not the most critical.	High; focused on both key improvements in finance security and efficiency	Moderate; technical nature may make it of interest to niche audiences only

legal troubles. Regarding environmental matters, energy consumption is a problem. The final security vulnerabilities here include the probability of being hacked or the 51% attacks that may compromise the reliability of the blockchain systems [5].

14.6.1 Limitations

Scalability Issues: Blockchain networks suffer from scalability issues, especially when the number of transactions increases inside the network. This translates to slower times in transactions and higher fees. This means that blockchain cannot be used on a large scale for financial activities.

Energy Consumption: Verification, as in the case of the line of proof protocols for work, can be extremely energy-consuming, and it tends to be an environmental concern. The costs for such high energy would dampen the potential implementation of this form of technology in areas which possess limited capacity to allocate resources or heavily regulate such environmental elements.

Security and Privacy: While blockchain technology indeed increases security in that information becomes decentralized, it is not hack-proof and is susceptible to attacks, mainly a 51% attack. There are also privacy issues in public blockchains because data about transactions tends to go public for anyone on the network [9].

Interoperability: Blockchains are not inherently interoperable, and blockchain solution fragmentation is likely to persist. Thus, the lack of standardization might be a challenge for financial institutions when it becomes necessary for them to put together several systems into one implementation

14.6.2 Future Scope

Scalability Solutions Improved: One can expect that the latest applications, such as layer2 solutions such as Lightning for Bitcoin or using a mechanism, Proof of Stake, should greatly improve the possibility of scalability without trading off security.

Regulatory Frameworks: The more complex regulations can give the financial sector that factor of certainty for legal purposes and trigger another wave in the adoption of blockchain technology. Governments and other regulatory bodies all over the world are now making initial attempts to come up with frameworks regarding taxation, security, and privacy [7].

Decentralized finance (DeFi) Integrations: DeFi or decentralized finance would change the entire face of the financial service industry altogether with this rising economic system. Here, transactions such as lending, insurance, trading, and so forth are executed without the involvement of a third party. The development of DeFi would nudge blockchain further towards bettering its immediate purpose of improving financial efficiency.

Cross-Industry Adoption: When blockchain interoperability increases, it will be adopted in health services, supply chain management, and government services apart from finance, which will strengthen the entire utility and integration of its financial systems.

14.7 CONCLUSION

The chapter "Blockchain Technology in Finance" elaborates on the transformativeness of blockchain technology within the financial sector. Blockchain, with its decentralized and immutable ledger, makes every financial transaction and operation much more secure, transparent, and efficient. Key innovations, including smart contracts, tokenization of assets, and DeFi, have transformed traditional financial practices, cutting costs and boosting liquidity, as well as giving way to totally new business models. The integration of blockchain in a number of financial processes—from the most basic payment and settlement arrangements to the thorny issues in compliance and risk management—outlines some of the deepest potentials it has for solving century-old problems in the industry. Capabilities such as creating tamper-proof audit trails, reducing operational complexities, and improving regulatory compliance come with great advantages for both financial institutions and clients. However, the wide adoption of blockchain in finance still poses difficulties in terms of regulations, technological limitations, and the need for industry-wide standardization. One will have to figure a way out of all such problems if one wants to unlock the full potential of blockchain in changing financial services. Looking ahead, a steady chain of innovations and pilot projects further explores and refines blockchain applications, continuing to drive further advancement in FinTech. As technology matures and regulatory frameworks continue to evolve, blockchain will stand cardinal in finance in the future with new opportunities for growth, efficiency, and transformation.

REFERENCES

1. Jadhav, K. (2024). Systematic literature review on innovations and risks in financial blockchain applications. Proceedings of the 3rd International MultiDisciplinary Conference: "Integrated Sciences and Technologies," IMDCIST 2023, Yola, Nigeria. European Alliance for Innovation.
2. Daidai, F., & Tamnine, L. (2024). Blockchain technology adoption and future trends in financial applications. In Blockchain Applications in Finance. Springer International Publishing.
3. Bogucharskov, A.V., Pokamestov, I.E., & Tropina, Z.N. (2024). Blockchain disruption in trade finance: A global perspective. Journal of Global Economic Review, 7(1), 250260.
4. Arnosti, N., & Weinberg, S. "Challenges of Decentralization in Blockchain Finance: The Road Ahead for Cryptocurrencies", *IEEE Transactions on Blockchain and FinTech*, 15(2), 2024, 98109.
5. Daidai, F., & Tamnine, L. "Blockchain Technology in Finance: A Literature Review", *Journal of Theoretical and Applied Electronic Commerce Research*, 18(1), 2023,125145.
6. Karthikeyan, D., Amutha, B., Mekata Devi, K., Singh, N., & Anglin Kuruba, P. "Hybrid Rater to Quantify and Measure the Severity of Infection and Spread of Infection in Muskmelon", *Bulletin of Electrical Engineering and Informatics (BEEI)*, 13(3), 2024, 1667–1675. https://doi.org/10.11591/eei.v13i3.5432.
7. Schröder, T., Atkinson, J., Biermann, F., & Elsässer, P. "Blockchain and Regenerative Finance: Charting a Path Toward Regeneration", *Frontiers in Environmental Science*, 11, 2023, 223245.
8. Bogusz, C.I., Laurell, C., & Sandström, C. "Tracking the Digital Evolution of Entrepreneurial Finance: The Interplay Between Crowdfunding, Blockchain Technologies, and Cryptocurrencies", *IEEE Transactions on Engineering Management*, 70(4), 2023, 10991110.
9. Kim, T., & Zhang, Y. "Blockchain Applications in Decentralized Finance: A Comprehensive Review", *Journal of Banking & Finance*, 152, 2023, 106125.
10. Coyne, J. G., & McMickle, P. L. "Can Blockchains Serve an Accounting Purpose?", *Journal of Emerging Technologies in Accounting*, 14(2), 2017, 101–111.

11. Hashim, F., Shuaib, K., & Sallabi, F. "Connected Blockchain Federations for Sharing Electronic Health Records", *Cryptography*, 6(3), 2022, 47.
12. KhMadhloom, J. "Dynamic Cryptography Integrated Secured Decentralized Applications with Blockchain Programming", *Wasit Journal of Computer and Mathematics Science*, 1(2), 2022, 21–33.
13. Li, M., & Xiao, P. "Network Blockchain Security Sharing Model Based on Fuzzy Logic", *Computational Intelligence and Neuroscience*, 2022, 2022, 1509000.
14. Lykidis, I., Drosatos, G., & Rantos, K. "The Use of Blockchain Technology in eGovernment Services", *Computers*, 10(12), 2021, 168.
15. Ansari, M.A., & Singh, D.K. "Human Detection Techniques for Real Time Surveillance: A Comprehensive Survey", *Multimedia Tools and Applications*, 80, 2021, 8759–8808. https://doi.org/10.1007/s11042-020-10103-4.
16. Pincheira, M., Vecchio, M., & Giaffreda, R. Characterization and Costs of Integrating Blockchain and IoT for AgriFood Traceability Systems. Systems, 10(3), 57, 2022a.
17. Dewang, R.K., Yadav, M.P., Awasthi, S., Raj, O., Mewada, A., & Bawankule, K.L. "Data Secure Application: An Application That Allows Developers to Store User Data Securely Using Blockchain and IPFS", *Multimedia Tools and Applications*, 83(15), 2024, 45491–45517.
18. Saari, A., Junnila, S., & Vimpari, J. Blockchain's Grand Promise for the Real Estate Sector: A Systematic Review. *Applied Sciences*, 12(23), 11940, 2022b.
19. Yeh, K.H., Yang, G.Y., Butpheng, C., Lee, L.F., & Liu, Y.H. "A Secure Interoperability Management Scheme for CrossBlockchain Transactions", *Symmetry*, 14(12), 2022, 2473.
20. Zhang, X., Xue, M., & Miao, X. "A Consensus Algorithm Based on Risk Assessment Model for Permissioned Blockchain", *Wireless Communications and Mobile Computing,* 2022, 2022, 1–21.

15 Harnessing Blockchain for Biotechnology

Current Applications, Challenges, and Future Opportunities

Asad Khan, Mohd. Aquib Ansari, Arvind Mewada, Greetta Pinheiro, and Amrendra Singh Yadav

15.1 INTRODUCTION

Blockchain [1–3], a decentralised and immutable ledger technology, has gained attention far beyond its initial application in cryptocurrencies like Bitcoin. By enabling secure, transparent, and tamper-resistant record-keeping, blockchain has the potential to transform various industries, including finance, supply chain management, and healthcare. The biotechnology sector has also begun exploring blockchain to solve some of its most pressing challenges in recent years. Biotechnology [4–7], a field that applies biological systems and organisms to develop products and technologies, is central to advancements in medicine, agriculture, and environmental sustainability.

Supply chains have indeed a very complicated structure, coupled with huge amounts of data generated and intellectual property (IP) protection in high demand. With biotechnology blossoming, managing and securing data, the integrity of research, and transparency along the supply chain also grow challenges. The inherent challenges push one into the adoption of blockchain technology, especially in biotechnology. As it is evident from the fact that biotechnology requires huge volumes of genomic data as well as clinical trial results or research findings, data management indeed becomes an overdue priority along with security matters. Figure 15.1 illustrates the blockchain's overall architecture for data-sharing systems in bioinformatics. Traditional systems often fail to ensure such countermeasures against data breaches, unauthorised access, and manipulation of research data.

Blockchain technology indeed offers an impressive way out by providing such a decentralised, secure, and immutable record of transactions. In addition, the biotechnology industry encounters great challenges in IP management. The patenting processes of biotechnological innovations are often contested and complex, leading to litigation and disputes among the stakeholders. In this regard, blockchain can

DOI: 10.1201/9781003545620-19

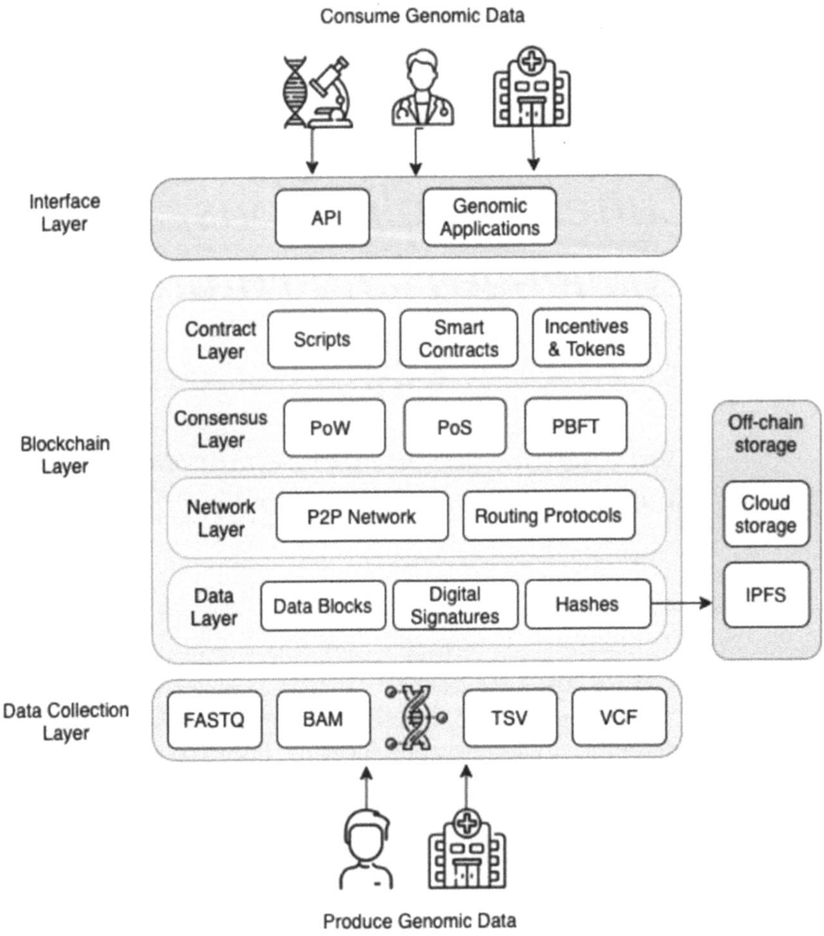

FIGURE 15.1 Blockchain architecture for data-sharing systems in bioinformatics.

ease IP management by providing a transparent and immutable record of innovations, enabling the tracking and management of patents and licenses granted to the record owner.

Ultimately, the biotechnology retail chain, which comprises different agents including manufacturers, analysts, controllers, and, at last, retailers, has emerged as a prominent player. An important feature of these supply chains usually guarantees their security along with quality, hence remaining highly transparent and traceable throughout. In addition to preventing counterfeiting and adhering to regulatory requirements, the capacity of blockchain to provide a verifiable record of each step in the supply chain can contribute to the resolution of these issues. This review chapter aims to provide a far-reaching outline of the ongoing biotechnology applications for blockchain innovation.

In this chapter, Section 15.2 provides a synopsis of the principal thoughts behind blockchain innovation, including its most significant highlights and types. In section 3, the utilisation of blockchain in biotechnology, for example, information the board, protected innovation the executives, store network straightforwardness, and customised medication, are analysed. Section 15.4 discusses the troubles, impediments, and specialised, administrative, and reception hardships of carrying out blockchain in biotechnology. Section 15.5 comprehensively examines existing arrangements, including various blockchain stages and a review of contextual studies. Section 15.6 frames open doors and future exploration headings for blockchain and biotechnology combination. In the last section, the chapter comes to a close with a summary of the most significant findings and some concluding thoughts on the significance of blockchain in biotechnology in the future.

15.2 BLOCKCHAIN FUNDAMENTALS

Most of the time, blockchain technology [8] is a distributed record framework that records transactions between various organisational computers. While traditional databases store data in a centralised location, blockchain operates on a decentralised network. All participants (nodes) have instant access to the same data as a result of this. This decentralisation is necessary for the transparency and security of the blockchain. The blockchain works by recording every transaction in a "block" linked to the previous block. Thus, a chain of blocks is formed, which is called a "blockchain." Data in these blocks is secured by cryptographic algorithms, thereby ensuring that once a transaction is recorded, it is unchangeable. The consensus mechanisms vary depending on the design of the specific blockchain, which controls the addition of a new block to the chain. The consensus algorithm is the process of agreement by all participants in a blockchain network concerning the state of the distributed ledger. The most used consensus mechanism among these is Proof of Work (PoW). Nodes solve difficult algorithms for determining the rightful owner of that data. The first node that solves the problem is given the right to append a new block and is awarded with cryptocurrency. It, however, does involve a lot of energy consumption. Typically, the Proof of Stake (PoS) is considered more efficient than PoW, as a node's probability of successfully adding a new block correlates with the number of tokens under its control. This work is said to consume minimal energy and is considered highly scalable.

Furthermore, Delegated Proof of Stakes enhances operation substantially since token holders get to nominate a certain cadre of delegates who will take responsibility for authenticating transactions and adding new blocks. Much quicker than PoW and PoS, the Delegated Proof of Stakes process has additional benefits. It applies mechanisms such as Byzantine Fault Tolerance to ensure that a blockchain can successfully operate, even with nodes resisted or erratic behaviour of nodes. This property is particularly important for private or consortium-like blockchains. In general, the features of cryptography, decentralised databases, transparency, and stability render blockchain technology superbly ideal for the industry's areas where credit is valued and where data integrity and security are most important, being biotechnology. The fundamental principles of cryptography, decentralised databases, transparency, and structural foundation make blockchain especially appropriate for industries where the integrity of credit is paramount, together with data authenticity and security, such as biotechnology.

15.2.1 Blockchain Types and Platforms

Three main categories of blockchain technology are mostly classified as public, private, and consortium. Examples of public blockchains include Bitcoin and Ethereum. These blockchains are open to everyone and completely decentralised. These types of blockchains may serve a significant role in biotechnology because their characteristics are critical for the transparent verification of information, especially when it comes to data from clinical trials. However, private blockchains are restricted to only one entity or a collection of organisations. They are mostly found in industries like banking and those that rely on supply chains, where privacy and efficiency are issues. In biotechnology, these private blockchains can be used to store sensitive patient information or move proprietary research data securely. Consortium blockchains are halfway decentralised and managed by a group of institutions and not by a single institution. Thus, they greatly balance openness and privacy and are pertinent to addressing collaboration challenges in biotechnology, like multi-centre clinical trials and consortium-based research initiatives.

Blockchain platforms have emerged to cater to specific needs in the biotechnology sector. For instance, Ethereum is known for its smart contract capability and is used to build decentralised applications across multiple industries, including biotechnology. It thus has the agility to automate even complex processes such as IP management and clinical trial workflows. Hyperledger Fabric is a modular, permissioned blockchain platform developed by the Linux Foundation; its current adoption in enterprises is due to factors such as scalability, privacy features, and plug-and-play components. Hyperledger Fabric can facilitate secure data sharing and supply chain management in biotechnology. Corda, initially tailored for the financial sector, stands out as a permissioned blockchain platform that prioritises interoperability and privacy. It is particularly adept at handling scenarios where multiple organisations need to share data without exposing it to all participants. This unique architecture makes Corda an invaluable tool in the biotechnology field, especially when it comes to managing contracts, safeguarding intellectual properties, and ensuring regulatory compliance.

15.2.2 Key Characteristics Relevant to Biotechnology

Some unique properties of blockchain technology make it an essential application in biotechnology. Above all, it immutably means that once data in a blockchain have been entered, they cannot be altered in any way. This peculiarity is necessary for maintaining data integrity in the biotechnology preservation of clinical trials and research, as well as in IP/documentation. Another very important advantage is a high degree of transparency; implementation of a blockchain system makes it possible for all the stakeholders in an organisation to be able to access the same information, thus increasing transparency and diminishing possible errors or fraudulent conduct. This is highly relevant in both clinical trials and supply chain management contexts. A prominent feature of blockchain technology is its decentralised quality.

It wipes out vulnerability due to weak links and eliminates single-point control, all by eliminating central authority. This feature proves to be very beneficial in global supply chain and collaborative research environments. The other characteristic of blockchain technology is that it is highly secure since the cryptographic methods used by blockchain technology are resistant to data tampering as well as unauthorised access. Blockchain technology, in the field of biotechnology, where data protection is of utmost importance, serves as a viable approach to protecting sensitive data. Furthermore, self-executing agreements with terms directly encoded into programming code are termed smart contracts, which offer great advantages. Smart contracts minimise dependence on intermediaries, increase efficiency in operation, and enforce agreements without requiring third-party involvement. This can be useful in the biotechnology processes of licensing IP, management of clinical trials, and supply chain transactions.

15.3 APPLICATIONS OF BLOCKCHAIN IN BIOTECHNOLOGY

15.3.1 Supply Chain Management

This indicates that a supply chain for biotechnology benefits extensively from using blockchain in that it provides an open, timestamped record of transactions. Typical members of a biotechnological item manufacturing network include suppliers of unrefined substances, manufacturers, regulators, and distributors. Products must be traceable and authentic in the pharmaceutical industry because fake drugs will have dreadful implications. Modern studies have proved that blockchain technology can indeed increase the productive efficiency of production networks through its inalterable record of all supply chain transactions. One good example is the MediLedger project [9], which, by using blockchain, is tracing pharmaceutical drugs such that only genuine ones reach the end-users. Further research has been conducted using blockchain technology in agriculture, tracing and ensuring that the origin and integrity of Genetically Modified Organism (GMOs), among other biotechnology products, are verified.

The utilisation of VeChain [10] in the pharmaceutical industry is an outstanding application of blockchain technology in the biotechnology production network. VeChain's blockchain platform enables pharmaceutical companies to track the entire lifecycle of drugs, from production to distribution. This prevents counterfeit goods from being distributed and ensures that regulatory standards are adhered to. The incorporation of blockchain with existing inventory network frameworks and the versatility of blockchain arrangements remain obstructions, even though blockchain offers critical benefits for inventory networks. Future assessments could explore hybrid models that unite blockchain with various advances, such as the Internet of Things, to further develop the creation of network-detectable quality and capability.

15.3.2 Data Management and Security

The biotechnology sector generates vast data, from genomic sequences [11] to clinical trial results. Managing and securing this data is a significant challenge, particularly given the sensitive nature of this information. Blockchain's decentralised and secure architecture offers a promising solution for data management in biotechnology.

Several studies have explored the use of blockchain for managing and securing biomedical data. For example, blockchain has been proposed to manage genomic data, where privacy and data ownership are critical concerns.

Projects such as the Personal Genome Project have experimented with using blockchain to give individuals control over their genomic data while ensuring it remains secure and private. Another project that is based on blockchain technology is the Iryo Network [12], which is an example of a platform that can implement the management of healthcare data. Through blockchain implementation, Iryo gives patients full control over their data—only individuals with appropriate rights can read the patient's records. It can be generalised to manage genomic and other sensitive biotechnological information. However, as with most solutions, blockchain is not without problems and issues, including the problem of storage and issues of compliance. Further studies might identify how such solutions can be integrated into large biotechnological systems and how proper legal regulations supporting blockchain technology in managing big data can be implemented.

15.3.3 Intellectual Property and Licensing

A well-protected area of a biotechnological business is mainly based on an IP measure since the growth of genetics, pharmaceuticals, and other related ideas can only be protected through patents and licenses. IP administration and protection is not an easy thing since it requires the assistance of many parties, as well as compliance with several different legal systems. Applying the principles of blockchain to the realisation of IP in biotechnology can be an effective method of enhancing the security of the procedures and the easy record of ownership and other transactions in the field [11]. Some investigations revealed that blockchain technology is efficient and can be applied to IP rights such as patents, copyrights, and licenses. The idea of IP transactions captured on a blockchain will help reduce theft incidences and help track licenses and enforce them.

Other examples implemented to leverage blockchain include the Intellectual Property Office of Singapore (IPOS), which shows how blockchain could bring more efficiency to IP processes. The poet strategy [13] is an example of a blockchain-based solution for controlling computerised content freedoms. Even though Poet's strategy was initially designed with digital media in mind, it could be applied to the management of IP in biotechnology to ensure that patents and licenses are transparently recorded and enforced. The intricacy of patent regulation, which changes fundamentally from one ward to another, is one of the principal hindrances to utilising blockchain for IP on the board. Blockchain, combined with existing patent and permitting frameworks and the production of normalised blockchain structures for IP boards, could be the subject of future exploration.

15.3.4 Clinical Trials and Research Integrity

Clinical preliminary studies are a fundamental component of biotechnology, particularly in developing novel drugs and treatments. However, concerns regarding the integrity of clinical trials are frequently sparked by issues such as manipulation of data, lack of transparency, and conflicts of interest. The unchanging and straightforward nature of blockchain makes it an ideal solution for ensuring the trustworthiness of examination and clinical preliminary information. The effectiveness of clinical trials has been the topic of several studies that have suggested using blockchain technology. For instance, Benchoufi and Ravaud [14] conclude that blockchain can timestamp and check each step of the clinical trial, which means that blockchain ensures data integrity and proper trial conduct. Another characteristic can also be traced back to the manipulation of trial results, and another key issue blockchain will assist pharmaceuticals with.

ClinTrial Token is an instance of how blockchain technology can be used to solve the problems of clinical trials. Since trial data is stored on a blockchain through ClinTrial Token, the data is reliable, transparent, and conform to the set standards. Overall, such an approach may go a long way in enhancing the reliability of trials, which in return enhances the investors' certainty in the pharmaceutical business. Looking at the advantages of blockchain linked with clinical trial transparency, there are still issues, such as data privacy and the problem of its implementation across the industry. For future works, it could be possible to propose studies linking further transparency and privacy that could be implemented with blockchain and studies that could better incentivise blockchain usage in clinical trials.

15.3.5 Personalised Medicine and Genomics

One rapidly developing field of biotechnology is personalised medicine, which focuses on targeted therapies considering a patient's genetic makeup. However, there are significant challenges related to the secure management and sharing of genetic data. Blockchain technology offers a promising solution specifically designed to protect sensitive information and facilitate the secure sharing of large data sets, such as genomic data. Multiple initiatives have been implemented in the application of blockchain in the field of Ph 4400. For instance, EncrypGen [15] uses blockchain to enable human beings to contribute markers of their genetics with researchers safely and of their own volition. Likewise, Nebula Genomics utilises blockchain to share genomics data, which people can sell to researchers or pharma companies.

The Shivom project is the blockchain technology system for genomic data sharing. Through blockchain, Shivom enables customers to share genomic data with researchers and pharmaceutical companies and simultaneously decide who should be able to use the information. What makes this approach beneficial is the possibility of propelling the research in personalised medicine without violating the individual's rights to privacy. However, one of the major issues that need to be addressed when applying blockchain in the context of personal medicine is data protection on the one hand and the ability to share data for research purposes on the other. Further study can be done in the direction of adopting blockchain with privacy techniques so that the data can be shared securely without compromising the privacy of the individuals. However, it is also known that the applicability of blockchain solutions for handling massive genomic data is one of the essential concerns for the future.

15.4 CHALLENGES AND LIMITATIONS

15.4.1 Technical Challenges

Identifying the benefits that blockchain brings to the field of biotechnology is crucial. Researchers also encounter numerous technical issues in using blockchain technology, which include issues on scalability, interoperability, and energy use. As of now, scalability remains the most prominent technical problem in blockchain. Even public blockchains such as Bitcoin and Ethereum can only process a number of transactions in a certain amount of time, leading to wait times and high fees per transaction. This limitation is most pronounced for practical applications in biotechnology that need huge amounts of computation or involve thousands of transactions. Another major problem with blockchain technology is interoperability.

Data sharing and collaboration, which are imperative concepts in the field of biotechnology, suffer due to the inability of different blockchain systems to cooperate. For instance, data transfer between different types of blockchains is usually not possible, and hence, the application of blockchain technology is complex. Finally, energy consumption is one of the significant challenges as it severely hampers the functioning of PoW-based blockchain networks. The heavy energy requirement for the functioning of PoW-based blockchain networks has raised concerns about the environmental impact and sustainability of such systems over time. This is particularly relevant for industries like biotechnology, which is increasingly looking towards sustainable approaches.

15.4.2 Regulatory and Ethical Issues

Although there is a trend of prohibiting certain uses of blockchain, particularly for biotechnology-related applications, it is still tightly regulated, and various issues will have to be settled before they can be utilised. It matters that legal uncertainty threatens the whole application area, especially regarding the situation with blockchain-based transactions or smart contracts applied by it. It may particularly act as a deterrent in highly regulated industries such as biotechnology, where the stakes are really high. Data privacy is another major challenge, especially when it comes to the use of big data analytics.

The immutability of blockchain records raises ownership concerns over data. Again, in biotechnology, where sensitive data, especially genomic data, are collected, compliance with data privacy laws such as the General Data Protection Regulation becomes very important. Moreover, there are also ethical issues at this point in using blockchain technology in biotechnology. For instance, issues of data ownership and consent come into play. For example, personalised medicine raises questions about data ownership and the potential exploitation of genomic information for economic purposes, where sensitive data commercialisation goes beyond its ethical limits. These are ethical concerns that must be addressed to ensure that this technology is responsibly utilised within biotechnology.

15.4.3 Adoption and Implementation Challenges

Many issues have to be dealt with and bridged to catalyse the adoption of blockchain technology in the biotechnology industry and its application. Resistance to change will be the biggest challenge that the adoption of blockchain technology will face. Most of the stakeholders here have spent decades within organisations relying much on conventional technologies, making it hard for them to shift to blockchain systems. High expense or sometimes even exorbitantly extravagant cost is a considerable barrier. The cost of implementing and maintaining the solutions using blockchain is really extremely high.

The financial advantages might accrue from its use over time, but in the end, investment in development and integration would be extremely limited due to its short-term understanding. Complexity does not facilitate adoption into blockchain as well. The chains in technology are complex, while the skills required in developing blockchain solutions are very specialised. This can be a big barrier for organisations without the relevant human resource profile because they cannot adopt and effectively use the technology.

15.5 COMPARATIVE ANALYSIS OF EXISTING SOLUTIONS

Based on the potential uses and varying characteristics of the blockchain in the biotechnology sector, various blockchain platforms are available. In this section, the authors describe the most promising blockchain solutions to be applied in biotechnology and compare them with advantages and disadvantages for specific tasks.

Ethereum vs. Hyperledger Fabric: Ethereum is very flexible and allows the support of smart contracts, making it suitable for use in IP management and clinical trial processes. It is a public platform, however, with scalability issues that make it less appropriate for dealing with sensitive data. On the other hand, Hyperledger Fabric is an open-source Blockchain Consortium (BCC) that is privacy-enhancing and flexible to use while implementing smart contracts; this suits large-scale organisations that use it in a supply chain and secure information exchange.

Corda vs Quorum: Corda is permissioned and also open, and therefore, it is well suited for establishing platforms through which different organisations can safely share data. Like Quorum, which is based on Ethereum but has granted access only to certain participants, the provision of anonymity in Vital is combined with the possibility of using a smart contract. It would be highly appropriate for both platforms to handle the various contracts and comply with biotechnology regulations; however, the colour choice will depend on the needs of each application.

This section also discusses several real-world cases of the successful application of blockchain in biotechnology, as well as an assessment of the effectiveness of such approaches and their potential consequences.

MediLedger Project: One of the projects that prove blockchain improves the supply chain and prevents the distribution of counterfeit products is MediLedger, which employs blockchain in tracking pharmaceutical products. This, therefore, shows how the use of blockchain enhances the safety and compliance with the standards of biotechnological products.

EncrypGen: Blockchain-based applications such as EncrypGen, which is used to manage genomic data through the blockchain, have described how blockchain can help people control how their data and information are used and shared securely. These sentiments apprise the platform's effectiveness and the potential facilitated by blockchain to overhaul data in personalised medication.

IPOS Blockchain Patent Platform: The IPOS has implemented blockchain technology to manage patents. The danger of patent disputes has been eliminated, and the patenting process has been accelerated, proving the viability of the application of blockchain platforms for managing all forms of IP in biotechnology.

15.6 FUTURE DIRECTIONS AND RESEARCH OPPORTUNITIES

Integrating blockchain technology with other relatively modern technologies, artificial intelligence [16–20], the Internet of Things, and big data presents great opportunities for developing biotechnology. For instance, integrating blockchain with artificial intelligence can make treatment planning and patient data analysis more effective in personalised medicine. Likewise, integrating blockchain with the Internet of Things may improve accountability in the supply chain for biotechnological products by allowing for tracking services. Although progress has been made in applying blockchain technology to biotechnology, there are still a number of research obstacles. These include developing standard structures for blockchain reception in biotechnology, working on the adaptability and interoperability of blockchain arrangements, and developing protection-saving blockchain advancements. Blockchain could affect biotechnology by making it conceivable to oversee information in a safer and viable way, expanding the straightforwardness of supply chains and shielding protected innovation. Notwithstanding, using this potential will require progressing blockchain innovation examination, joint effort, and speculation.

15.7 CONCLUSION

Blockchain technology can address some of the most pressing issues in biotechnology, such as supply chain transparency, IP management, and data security. In any case, a couple of particular regulatory and gathering difficulties should be addressed to grasp this potential. The primary focus of subsequent research should be overcoming these obstacles and examining the emerging opportunities at the intersection of blockchain and biotechnology.

REFERENCES

1. Pandey, Muskan, and Barkha Singhal. "Blockchain technology in biomanufacturing: Current perspective and future challenges." In SK Hafizul Islam, Arup Kumar Pal, Debabrata Samanta, Siddhartha Bhattacharyya, (Eds.), *Blockchain Technology for Emerging Applications*. Massachusetts, USA: Academic Press, pp. 207–237, 2022.

2. Yadav, Amrendra Singh, Nikita Singh, and Dharmender Singh Kushwaha. "Evolution of Blockchain and consensus mechanisms & its real-world applications." *Multimedia Tools and Applications* 82.22 (2023): 34363–34408.
3. Azaria, Azaria, Ariel Ekblaw, Thiago Vieira, and Andrew Lippman. "MedRec: Using blockchain for medical data access and permission management." In *Proceedings of the 2nd International Conference on Open and Big Data (OBD)*, Vienna, Austria, pp. 25–30, Aug. 2016. https://doi.org/10.1109/OBD.2016.11.
4. Vashishth, Tarun Kumar, Vikas Sharma, Kewal Krishan Sharma, Pranshul Sethi, Tanu Chaudhary, Abhay Bhardwaj, "Future implications of blockchain for biomedical and healthcare." In Prasun Kumar, Aparna Kumari (Eds.), *Blockchain for Biomedical Research and Healthcare: Concept, Trends, and Future Implications.* Singapore: Springer Nature Singapore, pp. 367–404, 2024.
5. Najumnissa Jamal, D., R. Karthikha, and M. S. Affia Thabassum. "Prospective issues and challenges for adopting blockchain for biomedical and healthcare systems." In Prasun Kumar and Aparna Kumari (Eds.), *Blockchain for Biomedical Research and Healthcare: Concept, Trends, and Future Implications.* Singapore: Springer Nature Singapore, pp. 269–301, 2024.
6. Balakrishnan, Arpita, et al. "Emerging trends of blockchain in bioinformatics: A revolution in health care." In Rajiv Pandey, Sam Goundar, and Shahnaz Fatima (Eds.), *Distributed Computing to Blockchain: Architecture, Technology, and Applications*, London, UK: Elsevier Academic Press, pp. 389–404, 2023.
7. Racine, Valérie. "Can blockchain solve the dilemma of the ethics of genomic biobanks?" *Science and Engineering Ethics* 27.3 (2021): 35.
8. Rizzardi, Alessandra, Sabrina Sicari, and Alberto Coen-Porisini. "IoT-driven blockchain to manage the healthcare supply chain and protect medical records." *Future Generation Computer Systems* 161 (2024): 415–431
9. Teoh, Bryan Phern Chern. "Navigating the blockchain trilemma: A supply chain dilemma." Advanced Maritime Technologies and Applications: Chapters from the ICMAT 2021. Springer International Publishing, 2022.
10. Simon, Richard, and Sameek Roychowdhury. "Implementing personalised cancer genomics in clinical trials." *Nature Reviews Drug Discovery* 12.5 (2013): 358–369.
11. Iryo, Takamasa, and Michael J. Smith. "On the uniqueness of equilibrated dynamic traffic flow patterns in unidirectional networks." *Transportation Research Procedia* 23 (2017): 283–302.
12. Aslam, Tehreem, Ayesha Maqbool, Maham Akhtar, Alina Mirza, Muhammad Anees Khan, Wazir Zada Khan, and Shadab Alam. "Blockchain-based enhanced ERP transaction integrity architecture and PoET consensus." *Computers, Materials & Continua* 70.1 (2022): 1089–1109.
13. Benchoufi, Mehdi, and Philippe Ravaud. "Blockchain technology for improving clinical research quality." *Trials* 18.1 (2017): 1–5.
14. DeFrancesco, Laura, and Ariel Klevecz. "Your DNA broker." *Nat Biotechnol* 37.10 (2019): 842–847.
15. Singh, Dushyant Kumar, et al. "Human crowd detection for city-wide surveillance." *Procedia Computer Science* 171 (2020): 350–359.
16. Dewang, Rupesh Kumar, et al. "Data secure application: An application that allows developers to store user data securely using blockchain and IPFS." *Multimedia Tools and Applications* 83.15 (2024): 45491–45517.
17. Rusia, Mayank Kumar, Dushyant Kumar Singh, and Mohd Aquib Ansari. "A novel deep transfer learning-based approach for face pose estimation." *Cybernetics and Information Technologies* 24.2 (2024): 105–121.

18. Ahmad, Shahnawaz, Shabana Mehfuz, and Javed Beg. "An efficient and secure key management with the extended convolutional neural network for intrusion detection in cloud storage." *Concurrency and Computation: Practice and Experience* 35.23 (2023): e7806.
19. Singh, Satya Prakash, Mohd Aquib Ansari, and Lalit Kumar. "Analysis of Website in Web Data Mining using Weblog Expert Tool." 2023 *IEEE 12th International Conference on Communication Systems and Network Technologies* (CSNT). IEEE, 2023.
20. Ahmad, Shahnawaz, et al. "Deep learning models for cloud, edge, fog, and IoT computing paradigms: Survey, recent advances, and future directions." *Computer Science Review* 49 (2023): 100568

Section 5

Security and Cryptography

16 Blockchain and Artificial Intelligence in Financial Industry 4.0

Enter the World of Digital Transformation

Annu and Ravindra Tripathi

16.1 INTRODUCTION

Artificial intelligence (AI) is the ability of a computer or machine to mimic human intelligence and conduct. Trend analysis, stock price predictions, fraud detection, and routine work automation are all possible with AI in finance. In recent years, significant advances have been made in AI technology. Google Inc.'s Alpha-Go AI software defeated human Go players in 2016, placing machine learning at the top of the rankings. After that, machine learning, neural networks, and other terms became common in financial markets and stock exchanges. Technology like AI and finance Industry 4.0 are changing the financial services business. Programming computers to learn, make judgments, and recognise patterns from past lessons is called "artificial intelligence" (AI). However, financial Industry 4.0 refers to adopting modern technologies like AI, big data, and blockchain in financial services [1,2].

Blockchain technology has transformed financial transactions by enhancing security and speed [3]. Asset tokenisation, cross-border payments, and smart contracts are further blockchain finance applications. The transaction will determine the ideal blockchain for finance. Private blockchains are better for private transactions, while public blockchains are transparent. Transaction speed, security, and regulatory compliance are critical when choosing financial blockchain models. Financial activity type also affects this choice. Blockchain records transactions transparently and unchangeably, improving financial operations. Due to its decentralised nature, fraud and manipulation are considerably reduced [4,5]. Fintech companies use AI and its many applications to create digital trading platforms that allow low-income women and young and small businesses to access the mainstream financial market.

AI and blockchain in the finance Industry 4.0 are solving risk management, fraud detection, and portfolio optimisation problems. AI also helps financial institutions offer real-time financial data and 24/7 client service. AI and blockchain in the financial Industry 4.0 are launching a new age of innovation and growth in financial services that might improve efficiency, accuracy, and accessibility for businesses and customers.

DOI: 10.1201/9781003545620-21

16.1.1 Industries Revolution Background

Coal in 1765: Steam power and manufacturing mechanisation began in the 18th century, starting the First Industrial Revolution. Mechanised spinning wheels produced eight times as much thread as conventional ones in the same period. Steam was noted for its efficacy. Its industrial use was the most significant productivity boost. Without humans, steam engines could power weaving looms. Technology like the steamship and (100 years later) the steam-powered locomotive allowed people and goods to travel farther in less time, leading to further advances.

Gas in 1870: With the rise of electricity and the production line in the 19th century, the Second Industrial Revolution began. Henry Ford (1863–1947) invented mass manufacturing in a Chicago abattoir where pigs were hung from conveyor belts, and each butcher only slaughtered a piece. "Henry Ford (1863–1947) implemented these concepts into automotive manufacturing, transforming the industry." Previously, an entire car was assembled at one location; now, small batches on a conveyor belt are more efficient and cost-effective.

Electronics and Nuclear in 1969: The decade of the 1970s marked the beginning of the 20th century's Third Industrial Revolution with the use of computers and memory-programmable controls. Since these technologies were introduced, production can be automated without human interaction. Well-known examples are robots that run pre-programmed sequences without human interaction.

Internet and Renewable Energy in 2000: *"Industry 4.0"* refers to corporate use of information and communication technology, building on advancements from the Third Industrial Revolution. Using a network link, "an online digital twin can be established and expanded for computer-based manufacturing processes." These allow facilities to communicate with one another and transmit information about themselves [6, 3]. Smart factories are created by cyber-physical production systems that connect all systems and require minimal human interaction. When these enablers come together, Industry 4.0 could revolutionise manufacturing. Examples are machines that foresee problems and start repair operations or self-organised logistics that adjust to production variations. Figure 16.1 depicts the Industrial Revolution.

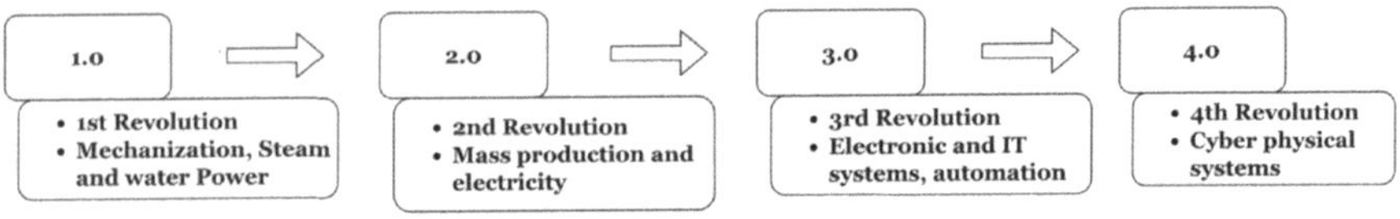

FIGURE 16.1 Industries revolution hierarchy.

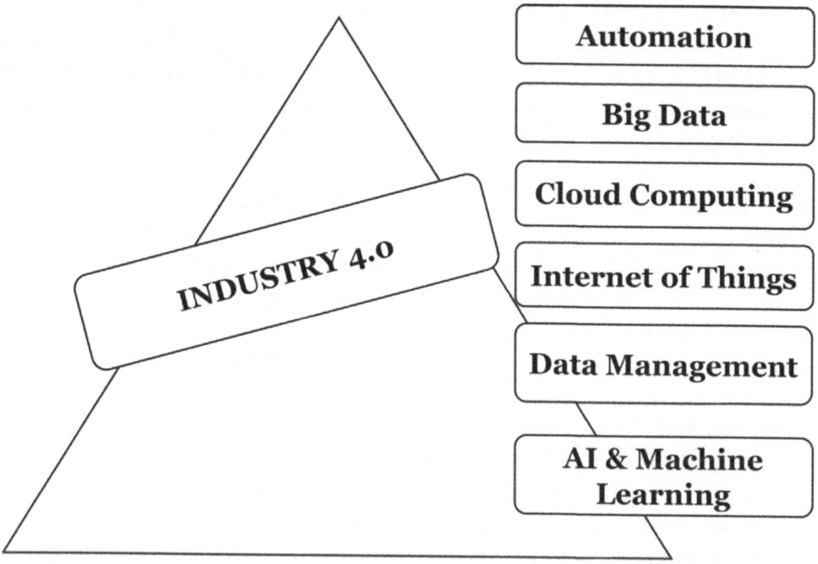

FIGURE 16.2 Involved technologies in Industry 4.0.

16.1.2 Involvement of Technologies

Figure 16.2 shows technological advances like cashless payments, digital payments, and Internet banking affect everyone daily. Investment technology is evolving, as shown by online trading apps. AI and machine learning improve digital market administration.

The stock market is complex and dynamic. Selecting an effective investment plan is tough. The challenge is accurate prediction. Stock prices are affected by politics, the market, and the economy. Investor decisions also affect them. Circumstances can significantly affect stock prices. Strict analysis makes finding a simple linear function equation for stock prices and other parameters difficult. Thus, while many basic investing models can recommend investments, they rarely foresee irregular financial goods. AI involves self-organisation, nonlinearity, learning, and self-adaptation.

16.2 THEORETICAL FRAMEWORK AND CONCEPTUAL MODEL

16.2.1 Artificial Intelligence and Stock Trading Industry 4.0 Relation

Modern financial markets can simplify market operations and increase trading strategy versatility. Financial institutions now operate better thanks to AI and machine learning [7]. From opening an online demat account to trading, everything is electronic. Technology improves usefulness and does not replace intelligence [8]. Financial markets today rely on trading software and transaction analysis instruments. Many trading systems allow online trading and demat accounts.

Stock Trading Industry 4.0 uses cutting-edge technology like AI to improve stock trading efficiency and efficacy. AI's purpose in Stock Trading Industry 4.0 is to analyse enormous amounts of data in real-time and make predictions and judgments. Additional trading solutions that employ AI to improve your trading experience are available. See how technology generates investor intelligence:

Risk Awareness: Businesses leverage machine learning, processing speed, and big data to analyse compliance risk-taking. This results from AI technology helping investors recognise complex trading patterns on a large scale. It also occurs in real-time across all marketplaces.

AI Stock Trading: Real-time financial notes, data, trending firms, and market information enable investors to trade. This is done via speech recognition and Natural Language Processing (NLP). Its primary benefit is saving traders time and effort.

Recommendations and Tips: Stock recommendations are made available through recognition technology, which helps investors make smart choices and maximise certain trading opportunities.

Less Human Error: because traders do not have to interact with people to get advice on trading tactics, etc. Data science quickly overcomes investment challenges.

16.2.1.1 Major Steps for the Formation of AI in the Financial Industry

Various steps, including AI in the financial industry, can be considered, as shown in Figure 16.3.

Data Collection and Preparation: The first step is to gather and process vast amounts of financial data to feed into AI algorithms.

Model Development: This involves selecting appropriate AI algorithms and developing models that can analyse financial data and make predictions or recommendations.

Training and Validation: The models have to be trained in big data so that accuracy improves; after that, these are validated to ensure effectiveness.

Integration and Deployment: AI models must be integrated with existing financial systems and deployed to end-users.

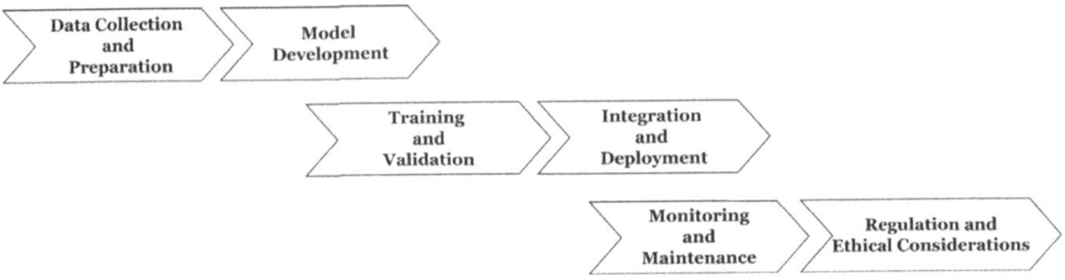

FIGURE 16.3 Formation steps of AI in the financial industry.

Monitoring and Maintenance: After their development, the AI models are monitored for performance and updated accordingly to ensure effectiveness.

Regulation and Ethical Considerations: The financial services industry is highly regulated. It should, therefore, develop and deploy AI in full consideration of the existing laws and regulations while exercising concern for ethical considerations.

16.2.2 Major Area of Financial Technology in Industry 4.0

FinTech represents one of the economy's fastest-growing sectors, using technology to improve and automate financial services. Major FinTech applications in Industry 4.0 include:

Mobile and Digital Banking: Greater penetration of mobiles and internet connectivity has also transformed the level of mobile banking and mobile merchant services.

Blockchain Technology: Blockchain technology increases financial transaction transparency, security, and efficiency.

Artificial Intelligence and Machine Learning: AI and machine learning will automate most finance processes regarding risk management, fraud detection, and the management of investments.

Robo-Advisors: Automated investment management services, often called robo-advisors, are becoming increasingly popular because they offer personalised investment advice to individuals.

Digital Currencies: Digital currencies such as Bitcoin and Ethereum are becoming more mainstream, leading to a growing interest in expanding new financial products and services based on blockchain technology.

Insurtech: FinTech companies are transforming the insurance sector by offering new products, services, and distribution channels, leveraging digital and data technologies.

16.2.2.1 How Has AI Made Its Mark in Financial Industry 4.0?

The financial industry's theoretical framework for AI is based on several key concepts and principles. One key concept is using machine learning algorithms that analyse and process large amounts of financial data, such as market trends, customer behaviour, and risk factors. Another important concept is using artificial neural networks to make predictions and decisions based on this data.

As seen in Figure 16.4, the conceptual model of AI in the financial sector usually consists of the following elements:

Data Processing and Analysis: The AI algorithms extract, clean, and process financial data for further use.

Machine Learning Algorithms: These algorithms use algorithms that have potential, such as decision trees, random forests, and neural networks, to analyse and make predictions on processed financial data.

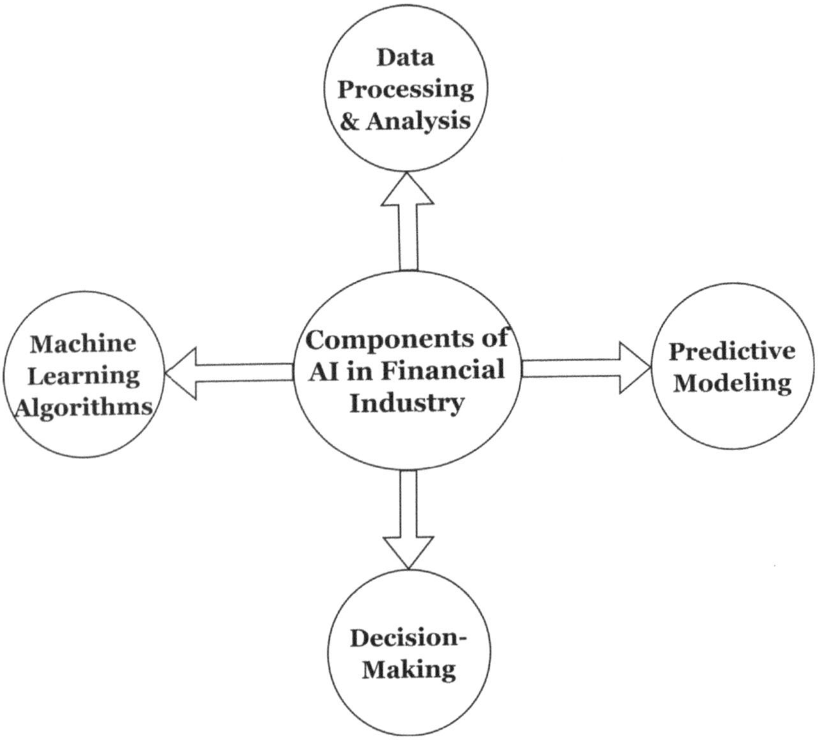

FIGURE 16.4 Components of AI in financial industry.

Predictive Modelling: This employed the output from machine learning algorithms to procure prognostications about the future trends in the marketplace, customers' behaviour, and risk factors.

Decision-Making: This implies the application of the predictive modelling module in decision-making processes, such as finding the best approach to investments or minimising risks.

Overall, the conceptual model of AI in the financial industry aims to use advanced technologies and algorithms to improve decision-making and enhance the efficiency of financial operations.

16.2.2.2 How Has Blockchain Technology Made Its Mark in Financial Industry 4.0?

Blockchain is essential and promising in Industry 4.0 [9,10], and this technology is believed to transform the world economy. It offers many chances for established businesses to prosper and new ones to start while challenging traditional enterprises. Blockchain is mainly connected with Bitcoin and other cryptocurrencies. Its benefits extend beyond cryptocurrency to many commercial and social transactions.

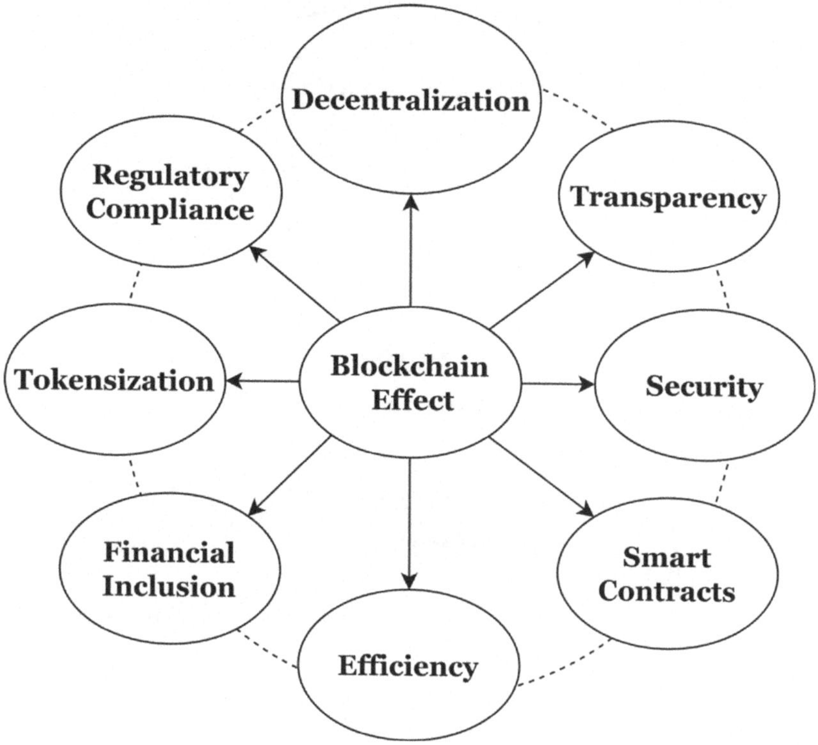

FIGURE 16.5 Several facets of blockchain in the financial sector.

Thus, blockchain could be used in different business applications [11,12]. Blockchain's strong cryptographic encryption, network participants' ongoing monitoring, and decentralised transaction data make it fraud-resistant. These features make manipulating system transactions nearly impossible [13,14].

Figure 16.5 illustrates how blockchain technology affects numerous facets of the financial industry by offering decentralised, transparent, safe, and efficient solutions.

16.3 THEORETICAL AND PRACTICAL IMPLICATIONS

16.3.1 Why Should I Feel Confident Investing in Artificial Intelligence?

Face recognition on phones, online shopping recommendations, finding the perfect job, a tailored music list, and Google Maps to handle intricate and unpredictable traffic are all examples of AI making our lives easier. AI affects us unknowingly [15]. AI-powered 'Jarvis Invest creates and manages a portfolio using 12 million financial criteria. We manage risk with this portfolio's 24×7 risk management. It can eliminate standard financial measuring methods' flaws in complex settings [16]. Indeed, quantitative investment using machine learning and AI has advanced swiftly and benefited investors.

FIGURE 16.6 Employment of AI in stock market Industry 4.0.

16.3.2 Applications and Challenges of AI in Stock Market Industry 4.0

AI is employed in the stock trading sector in various applications, as illustrated in Figure 16.6.

Portfolio Management: Using parameters like risk tolerance, investment horizon, and performance objectives, AI algorithms can be utilised to optimise the composition of a portfolio of investments.

Development of Trading Strategies: AI may be used to find relationships and patterns in massive data sets that can be utilised to create trading strategies. These tactics may be based on fundamental analysis, technical analysis, or a mix of the two.

Trade Execution: AI can automatically execute trades based on preset parameters or rules. Circumstances set in algorithms tracking markets and then executing trades when such circumstances are filled can be used.

Risk Management: With real-time monitoring of trades and portfolios, AI analyses and controls risks once they occur. Examples include detecting unusual trading patterns, noticing market conditions, and making any necessary changes in the trades based on them [5].

Market Analysis: With market data analysis, AI can provide insight for traders making the right decisions, like finding patterns and trends from market data, predicting price changes, or even giving real-time market analyses.

AI may increase equity trading efficiency, reduce cost, and enhance accuracy. However, AI is only the most accurate if its data and algorithms are the best; errors or biases in these will lead to errors [17]. In actively managed equity funds, AI advisers replace advisers when picking stocks. BlackRock is the most prominent US investment manager stock picker with a self-learning AI-based investment program. AI stock trading employs Robo-Advisors that scan for vast data and make a trade at optimal timing and price. AI traders are more visionary and effectively control a company's corporate trades, reducing risks and earning more returns [18,19]. Still, the financial sector faces many challenges in embracing AI technology. One of the main setbacks is the absence of technical people trained and skilled enough to set up AI systems.

The cost of AI is expensive in terms of purchase and large-scale implementation. Privacy issues also arise due to the immense amounts of data AI systems need to function effectively, posing security issues to financial institutions. More automation via AI also leads to job loss and, hence, the redeployment of workers. It is limited to the accuracy because also by the quality of data on which the AI systems have been trained that sometimes will convey biases in the resultant output. Many consumers still doubt AI in the financial sector, as they always distrusted an AI-powered system where they will keep storing their finances and other assets [20]. The deployment of an AI system is technically cumbersome as it has to interface with the existing processes. There is also a talent shortage in AI, making it difficult for financial institutions to find qualified professionals to develop and implement these systems. Lastly, the high costs associated with AI development and implementation can make it challenging for institutions to justify the investment.

Among the challenges that emerge from AI in finance are lack of transparency or bias by a financial AI system, data quality, regulation, expertise, risk management, and black box algorithms. Financial AI algorithms are complex, and an investor finds it challenging to evaluate AI-based investment strategies. Biassed AI systems could lead to the wrong financial output. AI algorithms generate forecasts based on the large amounts of data available and quality counts for precision [6]. It is challenging for investors to analyse AI-powered investment risks because they are not precisely controlled and regulated in finance. The topic of AI is expanding very fast, and many investors cannot appreciate the risks and benefits associated with AI-powered investment techniques. AI algorithms can process large amounts of data quickly, offering real-time financial choices that complicate risk management for investors. AI algorithms can indeed be "black box" systems, making investment decisions incomprehensible or impossible for investors to understand. The above issues thus underpin the importance of an informed investor being well-informed about the potential dangers and benefits of AI-powered investment strategies and the deliberative nature of the investment.

16.4 CONCLUSION, IMPLICATIONS, AND FUTURE RESEARCH DIRECTIONS

As a contribution to the finance literature, the paper further exposes an analysis of the role of AI within the financial industry as launched by the Fourth Industrial Revolution, Industry 4.0. It portrays the technological evolution from the first industrial revolution and draws insights into AI and stock trading as part of financial technology, or FinTech 4.0. The research shows how AI and blockchain change the financial sector by providing better safety, efficiency, and inclusivity at lower costs and without intermediaries. Given this pattern, technology can further impact the financial sector and provide growth opportunities.

Industry 5.0 would be the shift from technology-centric processes to human–machine collaboration. As such, Industry 5.0 would emphasise the integration of AI with human expertise, in turn, being able to produce highly personal, efficient, and innovative outputs (see Figure 16.7). Therefore, if manufacturing and financial services fail to keep pace with changes in the sector, Industry 5.0 will make them obsolete.

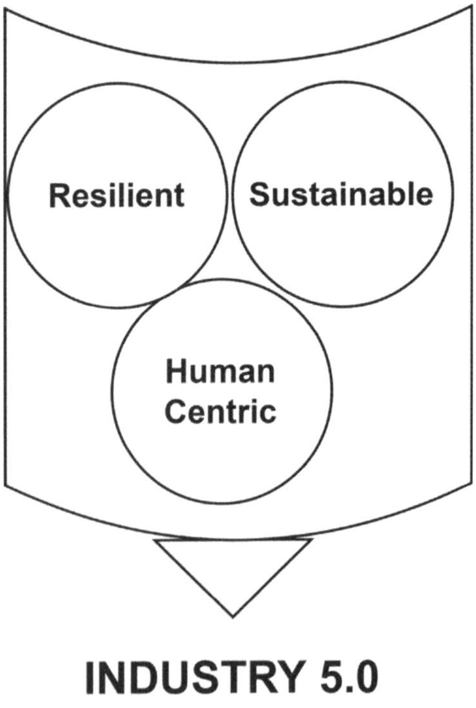

FIGURE 16.7 Core values of Industry 5.0.

While AI brings sundry advantages, high cost remains one of the main barriers. The costs of implementing AI on efficiency and privacy must be overcome so financial institutions can fully adopt and implement AI. Future directions should be to make technologies more accessible and affordable.

Moreover, future research should conduct surveys on investors to assess their knowledge about AI and how complete AI integration can impact investment choices. Further research into AI synergy with new technologies may lead to better decisions that benefit investors and businesses. AI can be fully utilised in the financial sector by addressing concerns over cost and privacy.

REFERENCES

1. Addula, S. R., Meduri, K., Nadella, G. S., and Gonaygunta, H. (2024). AI and blockchain in finance: Opportunities and challenges for the banking sector. *IJARCCE*, 13(2), 184–190.
2. Pal, A., Tiwari, C. K., and Behl, A. (2021). Blockchain technology in financial services: A comprehensive review of the literature. *Journal of Global Operations and Strategic Sourcing*, 14(1), 61–80.
3. Lui, A., and Lamb, G. W. (2018). Artificial intelligence and augmented intelligence collaboration: Regaining trust and confidence in the financial sector. *Information and Communications Technology Law*, 27(3), 267–283.
4. Rane, N., Choudhary, S., and Rane, J. (2023). Blockchain and Artificial Intelligence (AI) integration for revolutionising security and transparency in finance. *SSRN Electronic Journal*. 10.2139/ssrn.4644253.
5. Santos, H., Damasio, J., Ulbrich, A., and Vieira, R. (2021). Opportunities and challenges in fall risk management using EHRs and artificial intelligence: A systematic review. In *Proceedings of the 23rd International Conference on Enterprise Information Systems*. https://doi.org/10.5220/0010424306260633.
6. Kshetri, N. (2021). Blockchain and sustainable supply chain management in developing countries. *International Journal of Information Management*, 60, 102376.
7. Tadapaneni, N. R. (2020). Artificial Intelligence in Finance and Investments. *IEEE International Conference on System, Computation, Automation and Networking* (ICSCAN), Pondicherry, India, 2020, pp. 1–5.
8. Ghandour, A. (2021). Opportunities and challenges of artificial intelligence in banking: Systematic literature review. *TEM Journal*, 10(4), 1581–1587.
9. Morkunas, V. J., Paschen, J., and Boon, E. (2019). How blockchain technologies impact your business model. *Business Horizons*, 62(3), 295–306.
10. Viriyasitavat, W., and Hoonsopon, D. (2019). Blockchain characteristics and consensus in modern business processes. *Journal of Industrial Information Integration*, 13, 32–39.
11. Chang, V., Baudier, P., Zhang, H., Xu, Q., Zhang, J., and Arami, M. (2020). How blockchain can impact financial services – The overview, challenges and recommendations from expert interviewees. *Technological Forecasting and Social Change*, 158, 120166.
12. Hughes, A., Park, A., Kietzmann, J., and Archer-Brown, C. (2019). Beyond bitcoin: What blockchain and distributed ledger technologies mean for firms. *Business Horizons*, 62(3), 273–281.
13. Beck, R., Avital, M., Rossi, M., and Thatcher, J. B. (2017). Blockchain technology in business and information systems research. *Business, Information Systems Engineering*, 59(6), 381–384.
14. Boratyńska, K. (2019). Impact of digital transformation on value creation in Fintech services: An innovative approach. *Journal of Promotion Management*, 25(5), 631–639.

15. Chopra, R., and Sharma, G. D. 2021. Application of artificial intelligence in stock market forecasting: A critique, review, and research agenda. *Journal of Risk and Financial Management*, 14, 526.
16. Gao, Z. (2020). The application of artificial intelligence in stock investment. *Journal of Physics: Conference Series*, 1453(1), 012069. https://doi.org/10.1088/1742-6596/1453/1/012069.
17. Jan, Z., Ahamed, F., Mayer, W., Patel, N., Grossmann, G., Stumptner, M., and Kuusk, A. (2023). Artificial intelligence for industry 4.0: Systematic review of applications, challenges, and opportunities. *Expert Systems with Applications*, 216, 119456.
18. Wu, B., and Duan, T. (2019). The application of blockchain technology in financial markets. *Journal of Physics: Conference Series*, 1176, 042094.
19. Peres, R. S., Jia, X., Lee, J., Sun, K., Colombo, A. W., and Barata, J. (2020). Industrial artificial intelligence in industry 4.0 –Systematic review, challenges and outlook. *IEEE Access*, 8, 220121–220139. https://doi.org/10.1109/access.2020.3042874.
20. Murinde, V., Rizopoulos, E., and Zachariadis, M. (2022). The impact of the FinTech revolution on the future of banking: Opportunities and risks. *International Review of Financial Analysis*, 81, 102103.

17 International Payments Using Blockchain
A Comprehensive Analysis of Efficiency, Security, and Feasibility

Arvind Mewada, Mohd. Aquib Ansari, Niharika Keshari, Nagendra Singh, and Amrendra Singh Yadav

17.1 INTRODUCTION

With the rapid development of internet technology, electronic money transfers or international payments are growing rapidly. Nowadays, the international payment system plays an important role in performing global e-commerce activities. The total value of global international payments has reached USD 11.53 trillion in 2024 and is estimated to increase to USD 21.8 trillion by 2028, as per the World Bank statistics [1]. India is a major global trading nation that generates a large volume of international payments due to its broad offline import and export activities. In addition, the demand for international payments in India is increasing day by day due to factors such as e-commerce transactions, studying abroad, and tourism.

Nonetheless, the existing international payment model is suffering from challenges such as cumbersome processes and high transaction fees. Moreover, as these payment systems are exposed to serious security threats, these can expose user privacy as well as data protection risks [2]. Even this system has lengthy processing times and inefficient capital usage, which have further hampered the growth of e-commerce. Therefore, addressing these challenges and optimising international payment methods have become an important research area in the international payment system.

The cryptocurrency Bitcoin introduces the groundbreaking concept of blockchain technology. This technology has transformed financial applications, particularly in digital payment systems, both nationally and internationally. The greatest advantage of blockchain is that it has the ability to secure transactions and foster trust at a lower cost. This kind of feature of blockchain has garnered significant attention from government institutions, organisations, technology-based companies, and financial institutions worldwide [3]. This is known as the first generation of blockchain technology.

DOI: 10.1201/9781003545620-22

The second generation of blockchain technology introduced smart contracts, enabling functionalities beyond digital transaction transactions to assets such as stocks, loans, and real estate. The third generation of blockchain technology further expands its scope and offers solutions for industries ahead of banking and finance, such as healthcare, government, and even the science sector [4,5].

The National Institution for Transforming India (NITI Aayog) is a policy think tank for the government of India. NITI Aayog published a draft discussion paper in January 2020 titled "Blockchain: The India Strategy." In this report, the Aayog highlighted the journey or development of blockchain technology within the country (India) [6]. The Indian government has initiated various blockchain initiatives across sectors to promote the technology and related industries. For instance, in February 2020, the "Maharashtra Blockchain Sandbox" was launched, focusing on blockchain-based applications for e-governance, particularly in supply chain management, land registry, and public record-keeping [7]. The IndiaChain project, established in 2017, also aimed to create a blockchain-based platform for exchanging information and services among government agencies [8].

In the banking sector, ICICI Bank of India integrated blockchain technology into its Internet e-commerce financing system in August 2016. Furthermore, in December 2022, the Reserve Bank of India (RBI) launched the Digital Rupee, a central bank digital currency utilising blockchain technology. Unlike other cryptocurrencies, the Electronic Rupee (ER) has intrinsic value and is regulated by the RBI. The State Bank of India (SBI) Bank of Communications is also actively developing blockchain-based international payment application systems [9].

17.2 RELATED WORK

Blockchain is a digital distributed ledger that employs cryptographic techniques to secure and link digital records known as "blocks." These blocks are interconnected in sequential order through cryptographic "hashes," which form a secure chain of blocks. The foundation principle of blockchain has evolved since the 1980s but has been popularised worldwide after the advent of Bitcoin. Chaum [10] performed an early innovation in which he proposed blind signatures. It is a cryptographic method that is used for automated payments while ensuring the payee, transaction amount, and timing remain anonymous to third parties. Performing advancements in the area of cryptography, Haber and Stornetta [11] presented the concept of timestamping digital documents. It enhances security and authenticity by using timestamps embedded in records.

The author Finney [12] performed further extended advanced research in this field with the Reusable Proof of Work concept. This represents a groundwork for a secure and transparent digital currency payment system that could operate independently of traditional banks or central authorities. This research contribution shows a conclusion in the form of the first Bitcoin transaction performed in 2009. In this transaction, the researcher transferred ten Bitcoin (BTC) to a friend, and this marks the appearance of blockchain technology.

Bhaskar and Chuen [13] established the first Bitcoin exchange named "Bitcoin Market." It is marked a significant milestone for the blockchain development.

Buterin [14] follows this and introduces the concept known as a smart contract, which is integral to Ethereum based on a blockchain platform designed for decentralised applications and transactions beyond digital currencies.

Swan [15] performed a study based on the application scope of blockchain technology. This study highlighted the potential of blockchain in the financial sector. Further ahead, Patki and Sople [16] analysed the impact of blockchain on traditional financial institutions. They find the influence of blockchain on Indian commercial banks. Zhang [17] explored the application of blockchain in international payments systems. The author notices the potential of blockchain to create a globally integrated clearing system and to reduce transaction costs.

Deng [18] analysed the innovative potential of blockchain in international payment systems. He suggested strategies for blockchain trial and broader promotion. Similarly, Mori [19] performed a case study on Ripple. He demonstrated the advantages of blockchain in payment and settlement processes. Li [20] examined the limitations of existing traditional international payment models. He provides details on the benefits of blockchain-based payment systems and highlights their challenges.

Shen et al. [21] proposed an international payment system based on blockchain. They describe the advantages of blockchain in simplifying transactions. Roy et al. [22] compared various blockchain payment models used in international payments. Deng [23] also highlighted the potential of blockchain to address the bottlenecks in traditional international payment systems. Despite these advantages, Deng warned that risks associated with the implementation of blockchain technology based on international payments still need thorough exploration.

The literature of this chapter reveals that inefficiency, high costs, and security risks are the most prominent challenges associated with traditional international payments. This makes these issues a focus point for researchers in this field of research. The advent of blockchain offers promising solutions to these pain points using the introduction of decentralised, tamper-proof, anonymous, and traceable systems.

This chapter explores the potential of blockchain technology to overcome the limitations of existing international payment systems. Initially, the chapter theoretically examines the advantages of blockchain, which simplifies the payment process, enhances payment efficiency, reduces transaction costs, and ensures the security and authenticity of transactions. Then, designed a case study of an international payment system based on blockchain, which integrates real-world operations. Finally, the chapter presents recommendations for implementing blockchain in international payments, offering valuable insights for developing and commercialising blockchain-based payment systems in India.

17.3 BLOCKCHAIN TECHNOLOGY OVERVIEW

Blockchain synthesises distributed data storage, peer-to-peer (P2P) transmission, cryptographic algorithms, consensus mechanisms, timestamps, incentive structures (including rewards and punishments), and smart contracts. At its core, it is a decentralised, collectively maintained, and reliable ledger technology. Due to its decentralised, tamper-proof, anonymous, programmable, and secure nature, blockchain holds significant application value in areas such as digital currencies, international

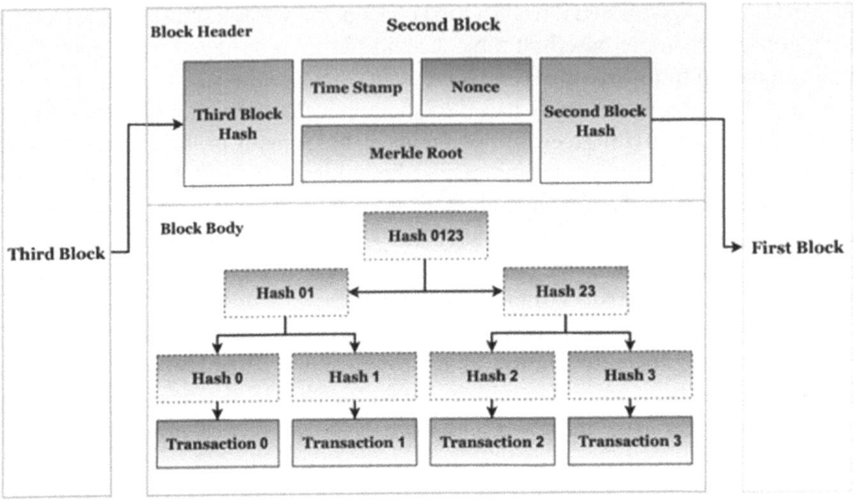

FIGURE 17.1 A block structure.

payments, and trade finance. Blockchain is composed of multiple blocks linked together to form a chain. Each block is structured with two main components, as depicted in Figure 17.1, representing a typical block structure: the block header and the block body.

The block body stores verified transactions, while the block header contains important metadata. This metadata includes the hash of the previous block, the hash of the current block, the timestamp, the nonce, and the Merkle root, all of which ensure the block's integrity and its connection to the rest of the chain.

The blockchain architecture is divided into six layers, as shown in Figure 17.2. Among them, the data layer is used to describe the physical form of the blockchain. It primarily includes data blocks, hash functions, timestamps, Merkle trees, chain structure, public-key encryption, etc. In the network layer, there are distributed networking, communication, and data verification mechanisms to distribute, forward, and verify blockchain transactions, etc. Typically modelled as a P2P network for all nodes that are equal participants. As soon as a transaction occurs, it is sent to nearby nodes, which check whether it is valid or invalid according to the rules set. Valid transactions are forwarded to further nodes, and invalid ones are discarded. In digital signatures, asymmetric cryptography is used to authenticate transactions through a signing and verification phase. It ensures secure interactions and communication among nodes in the blockchain network. The consensus layer contains various algorithms that are designed to achieve efficient agreement among untrustworthy nodes in decentralised environments. Blockchain networks do not have a trusted central node, so a rule (protocol) or process is needed to obtain consensus before a block is added to the blockchain network. So, in this layer, four major (Proof of Work (PoW), Proof of Stake, Practical Byzantine Fault Tolerance (PBFT), and Delegated

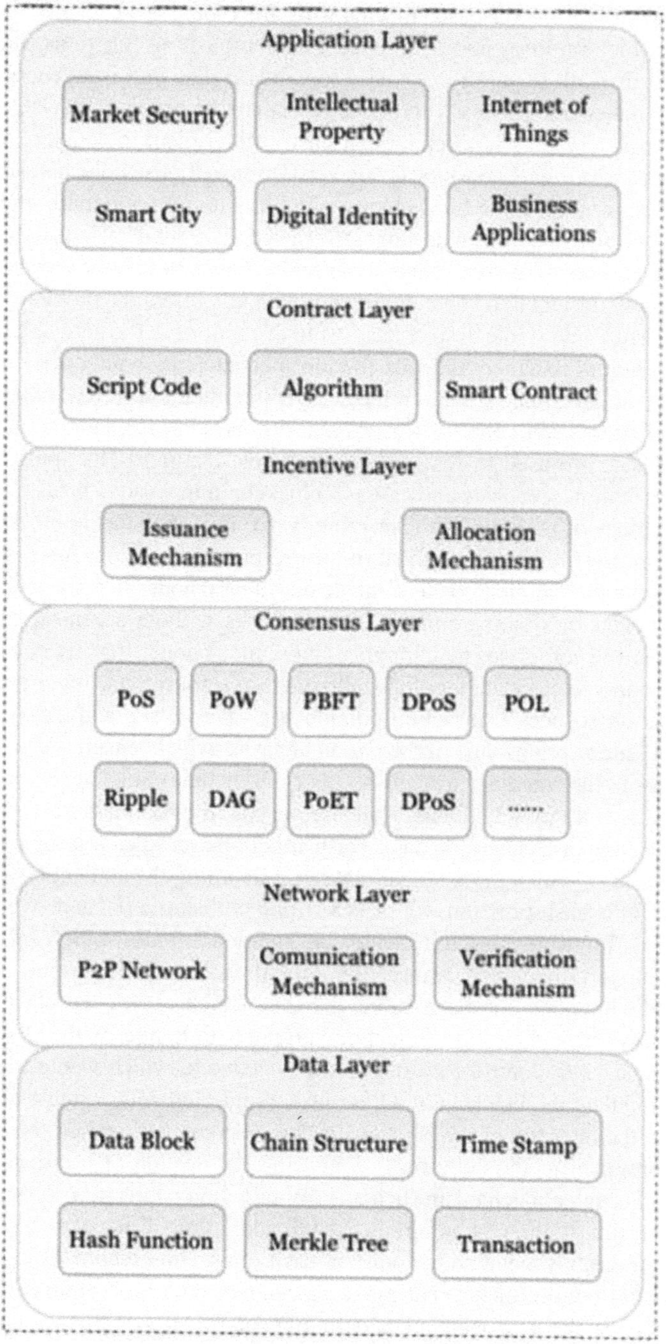

FIGURE 17.2 Blockchain layer architecture.

Proof of Stake (DPoS)) consensus mechanisms are widely used. The PoW is utilised in Bitcoin, which requires nodes to repeatedly run hashing functions to generate a nonce that is difficult to produce yet easy to validate. The PoS is an energy-efficient mechanism that enables nodes with the largest stakes to generate blocks.

The DPoS is a type of blockchain consensus mechanism in which token holders elect delegates who create new blocks and verify transactions on the blockchain. This system is designed to make blockchain operations more democratic and efficient. PBFT is a consensus algorithm designed to reach consistent decisions in blockchains and distributed networks even when some nodes behave erratically or maliciously. It is specifically designed to solve the Byzantine Fault Tolerance problem of keeping the system secure and stable despite fraud or malfunctions in the network. The incentive layer contains issuance and distribution mechanisms, which are mainly used to encourage nodes to participate in the security verification of blockchain transactions. The contract layer contains various algorithms, such as script codes and smart contracts. These are used to realise programmable features. The application layer includes application scenarios and cases of blockchain in various fields.

The processes of recording, transmitting, verifying, and storing blockchain data do not require the credit endorsement of a third-party institution. Instead, they rely solely on mathematical methods that enable distributed nodes to achieve a consensus of trust. Blockchain systems employ P2P networks without a central server. This network structure comprises many highly autonomous nodes that are equal in status and freely connected in a flat topology, allowing for resource sharing. In blockchain, decentralisation reduces transaction complexity, saves costs, and alleviates issues (connection interruptions and information delays), which ensures efficient value transfer without intermediary (organisation or middleman) links or servers.

Blockchain is a key technology that uses cryptographic methods to ensure the security and tamper-freeness of data. Hash functions serve as a unique tool under this security layer, which plays a vital role in maintaining the authenticity and integrity of data. This hash function works as a strong collision resistance, which is used to link blocks. It ensures that transaction information remains intact. An individual node attempts to tamper with transaction data already uploaded to the blockchain. So, when this happens, the hash value changes, which makes the block with that transaction invalid. Other nodes reject any attempts to tamper with the data unless someone manages to control 51% of the entire network, which would make it possible to manipulate the blockchain. Otherwise, tampering with a single node has no effect due to the integrity of the blockchain. Strong computational protection against attacks is provided to the authorisation and the verification of transactions in the blockchain through a mechanism such as PoW based on public key cryptography.

Moreover, timestamp technology is used to add a time dimension to data, which ensures the traceability and authenticity of each transaction record. Smart contracts work as digital commitments that make blockchain data programmable, which is used to enhance the management and trading of traditional financial assets and lay the foundation for programmable financial systems. They are characterised by decentralisation and autonomy. Decentralisation means that smart contracts are executed by decentralised stored and verified program codes, ensuring fairness and equity in fund transfers. Autonomy implies that once specific conditions are met, the contract executes automatically without external interference, ensuring the security and accuracy of transaction processing. By leveraging the programmability of blockchain smart contracts, complex centralised clearing processes can be eliminated, significantly reducing costs, improving efficiency, and enabling the quick and convenient trading of financial products.

Although transaction information in the blockchain is broadcast to the entire network, the identities of the transaction nodes remain anonymous. Each node does not need to trust one another, meaning identity information must not be publicly disclosed. This allows for anonymity, protecting user privacy. Each account identity in the blockchain system is represented by a wallet address that can be changed at will. This wallet address is generated through cryptographic technology, making it impossible to infer the user's identity or private key. No one can determine the identity of the transaction user corresponding to an account, ensuring that each transaction is carried out anonymously.

Blockchain is a distributed shared ledger maintained collectively by all nodes. The distributed data storage and transaction recording enable all users to participate in this bookkeeping system. The data on the ledger is public and transparent, allowing anyone to check, trace, and audit information through a public interface. This public transparency ensures the tamper-proof nature of blockchain data and allows for public oversight of compliance with operational behaviour. Consequently, the transaction information within the blockchain system is open and transparent, preventing one node from deceiving another. The transparency of blockchain promotes open and effective operations in international trade activities.

17.4 ADDRESSING CHALLENGES IN INTERNATIONAL PAYMENTS WITH BLOCKCHAIN TECHNOLOGY

Common international payment methods include bank wire transfers (as shown in Figure 17.3), international credit card payments, and third-party payment services, with bank wire transfers being the primary remittance option. These traditional payment settlements operate in a centralised architecture involving multiple intermediary entities. As a result, the payment-clearing process can be cumbersome,

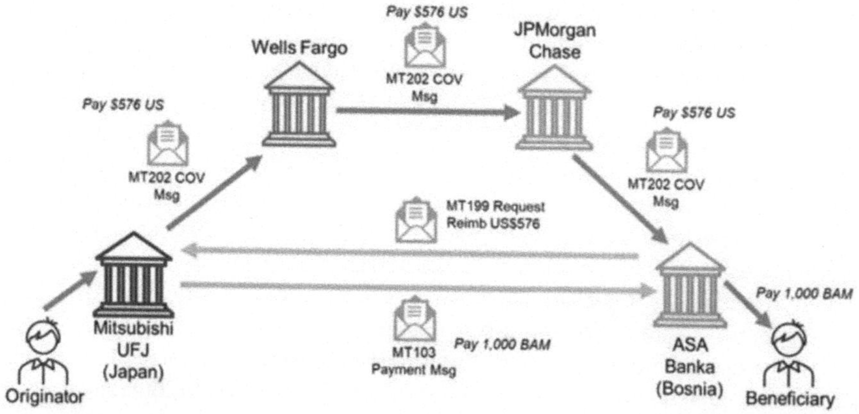

FIGURE 17.3 Wire transfer using society for Worldwide Interbank Financial Telecommunications.

leading to slow business processing speeds, low payment efficiency, and high costs. Furthermore, the lengthy and complex payment procedures place significant demands on bank personnel, and operational errors can have serious consequences for both parties involved in the trade. The emergence of blockchain technology offers potential solutions to the challenges of traditional international payments. Integrating blockchain into this process makes it possible to eliminate intermediaries, simplify settlement procedures, enhance payment efficiency, and reduce costs, making international payments faster and more cost-effective. Moreover, blockchain reduces risks in operations when performing international payments and protects user privacy.

17.4.1 Payment Efficiency in International Transactions

Payment efficiency mainly refers to processes and the associated time costs. Simplifying these transaction processes and reducing time costs is a very important point for international trade, as it ensures a smooth flow of money and helps reduce foreign exchange risks. Traditional wire transfer systems have multiple steps, such as remitting banks, correspondent banks, payment systems, and receiving banks. Each organisation or institution operates its own independent accounting system, which results in complex procedures such as fund recording, clearing, and reconciliation with the counterparty bank. Subsequently, this structure or framework contains lengthy processing times and low payment efficiency for international transactions. For example, a bank which performs wire transfers for international remittances generally takes 3–5 working days. However, international credit cards or third-party payment services take 3–7 days to process. In the case of large-scale transactions in which a manufacturer purchases raw materials from a supplier, the opportunity cost of waiting several days for payment clearance is significant. These delays can hinder production schedules, impact a company's ability to fulfil orders, and potentially jeopardise business relationships and financial stability.

Blockchain technology eliminates intermediatory mediators or organisations, reduces payment times in parallel, and maintains proper security. This allows remitters and recipients to make direct P2P payments and settlements. Furthermore, payments can be processed 24/7, regardless of holidays, enabling instant arrivals. In addition, withdrawals are convenient and free of hidden costs.

Stellar (XLM), Ripple (XRP), Corda, and IBM are blockchain platforms used worldwide to create a blockchain-based international payment system. Ripple is a popular platform due to its speed and low transaction costs, which facilitate real-time international transactions. Ripple is a decentralised consensus mechanism that has the ability to perform a transaction in short times to within 3 seconds, significantly improving settlement efficiency (see Figure 17.4).

The efficiency of international payments is of great significance to traders. In a competitive international market, fluctuations in foreign exchange rates significantly impact trading operations. International businesses reduce the exchange rate risk associated with long-term capital occupation while simultaneously accelerating capital turnover, contributing to market prosperity by shortening settlement times.

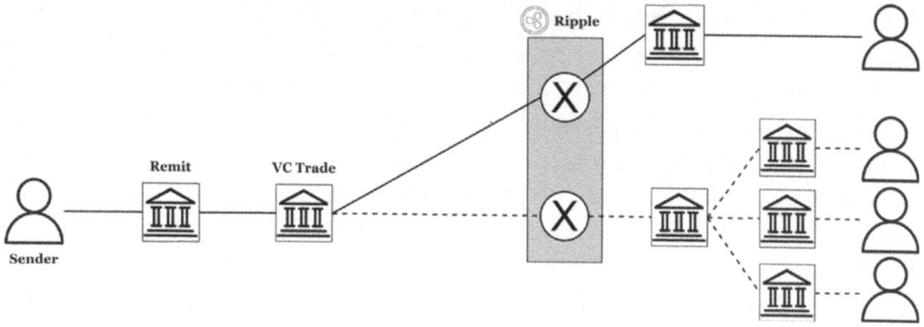

FIGURE 17.4 Ripple transfer system.

17.4.2 Cost Comparison of International Payment Systems

The traditional international payment system is very complex in India, and transaction costs are also quite high. The Indian banks charge a fee of approximately INR 500 to INR 1,500 for remittance amounts, along with additional charges such as telecommunications fees. For example, SBI or HDFC Bank charges transaction fees based on the remittance amount, which depends on the transaction size and destination. Subsequently, the total cost of each remittance falls in the range of INR 800 to INR 2,000.

Blockchain technology has the potential to significantly simplify the international payment process by reducing payment costs and saving transaction fees for traders. A McKinsey report assesses the introduction of blockchain technology in international payments. It globally decreases transaction costs from nearly $26 per transaction to $15 per transaction, which saves approximately 42% of the amount of both parties involved after the adoption of Ripple blockchain technology in an international payment system. A prominent Indian bank reduced fees from INR 400 per transaction to as low as INR 50. This technology saves users up to 87.5% in fees and greatly minimises payment costs.

Furthermore, a traditional international payment system requires trusted central banks to facilitate payments and settlements between different banks. They are also required to maintain separate margin accounts to minimise transaction risks. Due to the old payment method, a large amount of capital gets stuck for some time, and subsequently, the growth of foreign trade is adversely affected due to the low rate of utilisation of funds. This new method has particularly benefited small and medium-sized enterprises. However, platforms like Ripple eliminate the need for intermediary banks in a blockchain-based international payment system by enabling real-time payments. This approach requires only a single margin account, significantly reducing the amount of capital tied up and enhancing the utilisation rate of funds.

17.4.3 Enhancing Security and Trust in International Transactions

The inherent characteristics of blockchain technology ensure the security, transparency, and reduced risks associated with international payments. Blockchain-based payment methods operate on a push-based model, meaning that when a user sends a payment to a merchant, they do not need to provide their personal information or bank account details. Instead, the system uses a secure hash function to transmit a hashed value of the encrypted personal information to the merchant. Due to the one-way nature of the hash function, personal information remains protected and is not leaked. Furthermore, Ripple employs anonymous transactions, which offer additional privacy protection for users.

In contrast, traditional payment systems utilise a centralised pull-based method that requires real-name transactions. To complete a transaction, users must provide their personal information and bank account details to the merchant online. The merchant uses this information to withdraw funds. Consequently, the merchant accumulates many users' sensitive data, such as bank account numbers and other personal transaction details. This private information is vulnerable to being leaked or sold by the merchant and may also be at risk of theft by criminals, leading to substantial losses for users. Thus, adopting a blockchain-based payment system can enhance user privacy and improve the security of online transactions.

In addition, traditional international payments have significant risks due to relying on commercial credit without guarantees. The importers and exporters themselves must manage trade risks. One of the primary challenges is ensuring that the importer's payment and the exporter's shipping information are accurate and reliable. Once the importer remits the funds, they cannot track or control the intermediate links in the payment process, nor do they have visibility into the exporter's shipping status. The trading parties must rely solely on commercial credit for transactions, which leaves them vulnerable to commercial fraud. Even in the absence of fraud, poor business practices or sudden market changes can create bad faith, exacerbating these risks and leading to significant losses for traders. Such issues can erode trust in international trade and severely hinder its growth.

Blockchain technology can mitigate these trade risks in international payments. Participants in a transaction do not require the endorsement of intermediary institutions. Instead, they upload the transaction contract information to the blockchain, where all parties involved in the trade jointly maintain the transaction details and verify consistency. If one party does not confirm the transaction information, the transaction is rendered ineffective. The blockchain system facilitates automatic payments once both the importer and exporter validate the transaction information.

The payment process begins with the importer making a prepayment to the exporter per the contract. The exporter then ships the goods via a logistics company. Upon verifying the cargo information, the logistics company uploads the delivery note to the blockchain network. Once the importer confirms the transaction details on the blockchain, the remaining balance is settled. Therefore, blockchain-based international payments can effectively resolve trust crises, reduce payment risks, save time, expedite corporate funds' return, and promote the rapid and sustainable development of international trade.

17.5 CHALLENGES AND RECOMMENDATIONS FOR BLOCKCHAIN IMPLEMENTATION

The primary obstacle to implementing blockchain technology in international payment settlements is the inherent limitations of its core technology. Significant challenges still need to be addressed, including energy consumption, credit issues, security of encryption algorithms, block capacity, and lengthy transaction confirmation times. These technical hurdles impede the broader adoption of blockchain solutions. As a major trading nation, our country must enhance its efforts in independent research and development of blockchain core technology. Financial technology companies should particularly focus on the research outcomes of related domestic and international industries, investing more human and financial resources to actively develop key technologies and business models for blockchain in international payments. This proactive approach will enable us to establish technical reserves and position ourselves as leaders in the technology era.

The decentralised nature of blockchain presents a stark contrast to the current centralised financial model. Consequently, we incorporate blockchain technology into the international payment system, and it is crucial to establish a corresponding regulatory framework. As we know, no single country or institution independently regulates blockchain technology. Therefore, all financial institutions and regulatory authorities worldwide must collaborate together to formulate a blockchain regulatory strategy that suits global trade settlements. This technology ought to balance supervision, innovation, and practical application.

In order to establish a blockchain-based international payments system, we need to benefit from the UK-based regulatory sandbox mechanism. It allows pilot projects to be run across various sectors. For example, selecting blockchain for the transaction settlement banks within free trade zones for trial runs could provide valuable information. This approach enables researchers or developers to refine technology and regulatory policies before rolling out their larger-scale implementations.

Blockchain technology is inherently multidisciplinary, which requires expertise in mathematics, computer science, cybersecurity, communications engineering, and finance. Nevertheless, India presently requires more high-end talent with this composite skill set. To address this gap, leading universities in India are introducing specialised blockchain courses within their mathematics, computer science, or cybersecurity programs. It is recommended that reputable domestic and international blockchain experts from different organisations be engaged to support talent development. In addition, working together with financial technology companies helps prepare blockchain experts for international business transactions.

In addition, it is necessary to establish a uniform system in industry guidelines for implementing blockchain technology in international trade transactions.

The establishment of these standards fosters healthy competition among enterprises, which improves customer satisfaction. It enhances the effectiveness of financial markets and mitigates potential application risks. In addition, uniform industry standards are necessary to monitor the market and formulate better policies. For this, countries and financial institutions should agree on basic aspects such as blockchain terminology, coding, technical development, and standardisation of products and services.

17.6 CONCLUSION

A blockchain has the potential to enhance processing and reduce inefficiencies in international payments. Blockchain exists as a trusted and decentralised system free of intermediaries. The result is that the costs of transactions, processing time, and the security of financial transactions are also reduced. Blockchain uses cryptography to protect user's privacy and turn to reduce the risk of data breaches and fraud. In addition, blockchain is transparent and immutable; therefore, all transactions are secure and trackable. This also increases the trust in international payment systems, which is quite important when considering international trade and global e-commerce, where banks and payment systems must serve fast and secure payments. However, more research, collaboration, and investment are still needed to fully expose the true power of blockchain in international payment systems. The solutions to the problem of scalability and security in the blockchain space must be done in collaboration between governments and other financial institutions and technology companies to cater for global financial needs. It is a technical challenge that necessitates solving for energy efficiency, compliance with regulations, and a willingness to use innovation to push beyond how the blockchain can be used on a wide and global basis in international payments. But, as it turns out, blockchain is not just a technological change. It is a powerful force behind the future of international payments, bringing greater efficiency and security to the world's financial systems at a lower cost.

REFERENCES

1. N. Singh, H. P. Singh, A. Mishra, A. Khare, M. Swarnkar, and S. K. Almas. Blockchain cloud computing: Comparative study on DDoS, MITM and SQL injection attack. In *2024 IEEE International Conference on Big Data & Machine Learning (ICBDML)*, Bhopal, India, pp. 73–78, 2024. https://doi.org/10.1109/ICBDML60909.2024.10577412.
2. Wennan Wang, Linkai Zhu, and Shiyang Song. A study on privacy protection technology for cross-border transactions based on blockchain technology. In *Sixth International Conference on Computer Information Science and Application Technology (CISAT 2023)*, vol. 12800, pp. 567–572, 2023.
3. R. S. Veena, G. Khambra, R. Sayal, N. Singh, et al. An energy efficient IPv6 packet delivery outline for industrial IoT over G.9959 protocol based Wireless Sensor Network (WSN). *SN Computer Science*, 5:813, 2024. https://doi.org/10.1007/s42979-024-03195-4.
4. R. K. Dewang, M. P. Yadav, S. Awasthi, O. Raj, Arvind Mewada, and K. L. Bawankule. Data secure application: an application that allows developers to store user data securely using blockchain and IPFS. *Multimedia Tools and Applications*, 83(15):45491–45517, 2024.
5. Michael Crosby, Pradan Pattanayak, Sanjeev Verma, Vignesh Kalyanaraman, et al. Blockchain technology: Beyond bitcoin. *Applied Innovation*, 2(6–10):71, 2016.
6. Nagendra Singh and Yogendra Kumar. Security and privacy requirements in 5G healthcare. In *Block Chain for 5G Healthcare Applications*, pp. 1–30, Nov 2021. IET Publication. https://doi.org/10.1049/PBHE035E_ch1.
7. Kratika Narain, and & G. Doctor. Exploring blockchain-based government services in India. In Sanjay Vyas, Vipin Kumar Shukla, Sanjay Gupta, and Amit Prasad (Eds.), *Blockchain Technology: Exploring Opportunities, Challenges, and Applications*, pp. 127–144. Boca Raton, Florida: CRC Press, 2022.
8. Ajay Kumar Bharti, Anil Kumar, Pankaj Chauhan, Pradeep Jain. A study of emerging areas in adopting blockchain technology and its prospective challenges in India. In *2019 Women Institute of Technology Conference on Electrical and Computer Engineering (WITCON ECE)*, pp. 146–153. IEEE, 2019.
9. Amarpreet Singh Virdi and Akansha Mer. Fintech and banking: An Indian perspective. In Nader Naifar & Ahmed Elsayed (Eds.), *Green Finance Instruments, FinTech, and Investment Strategies: Sustainable Portfolio Management in the Post-COVID Era*, pp. 261–281. Cham, Switzerland: Springer, 2023.

10. David Chaum. Blind signatures for untraceable payments. In David Chaum, Ronald L. Rivest, and Alan T. Sherman, editors, *Advances in Cryptology*, pp. 199–203. Springer US, 1983.
11. Stuart Haber and W. Scott Stornetta. How to timestamp a digital document. Springer, 1991.
12. Hal Finney. RPOW reusable proofs of work. 2004. https://nakamotoinstitute.org/finney/rpow/.
13. Nirupama Devi Bhaskar and David Lee Kuo Chuen. Bitcoin exchanges. In *Handbook of Digital Currency: Bitcoin, Innovation, Financial Instruments, and Big Data*, pp. 537–551. Amsterdam, Netherlands: Elsevier, 2024.
14. Vitalik Buterin. A next-generation smart contract and decentralised application platform. *White Paper* 3(37), 2014.
15. M. Swan. *Decentralised Finance: Blockchains, Prediction, and Valuation*. Economist: Finance Disrupted, New York2016.
16. Aarti Patki and Vinod Sople. Indian banking sector: Blockchain implementation, challenges, and way forward. *Journal of Banking and Financial Technology*, 4(1):65–73, 2020.
17. Wei Xuan Zhang. The current situation and trend of blockchain technology in the financial field. *Mobile Information Systems*, 2022(1):7908625, 2022.
18. Qing Deng. Application analysis on blockchain technology in cross-border payment. In *5th International Conference on Financial Innovation and Economic Development (ICFIED 2020)*, pp. 287–295. Atlantis Press, 2020.
19. Taketoshi Mori. Financial technology: Blockchain and securities settlement. *Journal of Securities Operations & Custody*, 8(3):208–227, 2016.
20. Xiang Hong Li. Blockchain-based cross-border e-business payment model. In *2021 2nd International Conference on E-Commerce and Internet Technology (ECIT)*, pp. 67–73. Hangzhou, China. IEEE, 2021.
21. Xiang-Dong Shen, Xi Chen, Ran Ji, and Ren-Hong Wu. The new ecosystem of cross-border e-commerce in Korea, China, and Japan is based on blockchain. *Journal of Korea Trade*, 24(5):87–105, 2020.
22. Avishek Roy, Md Ahsan Habib, Shahid Hasan, and Abu Taher Md Shifat. A scalable cross-border payment system based on consortium blockchain ensuring auditability. In 6th International Conference on Electrical Information and Communication Technology (EICT), IEEE, 2023.
23. Qing Deng. Application analysis on blockchain technology in cross-border payment. In *Proceedings of the 5th International Conference on Financial Innovation and Economic Development* (ICFIED 2020), pp. 287–295. Atlantis Press, 2020. ISBN 978-94-6252923-6. https://doi.org/10.2991/aebmr.k.200306.050.

18 Blockchain-Driven Solutions for Energy Trading and Sustainability in Microgrids

Mohd Bilal, Ahsan Waseem, Imdadullah, Salman Hameed, and Zenab Elgamal

18.1 INTRODUCTION

The increase in global energy demand is directly proportional to urbanization and industrialization. Traditional energy systems use fossil fuel-based energy-generating sources, which cause a large emission of greenhouse gases. Rapid increases in demand can only be met by centralized power plants, which are environmentally friendly [1]. Generating power plants are situated far from the utilization zone, which causes large transmission losses, making the system less efficient. In many regions, the ageing infrastructure and inefficiency of the transmission system cause higher costs and environmental degradation.

Climate change and international agreements like the Paris Agreement emphasize developing a more renewable energy-based system. Renewable energy sources are intermittent, and their integration is a major concern for transitioning from a centralized energy system to a decentralized energy system. Renewable source-based energy generation depends on the geographical location, wind speed, availability of biomass fuel, and solar irradiance. Integration of these generating units effectively into the grid is a challenge. Secure energy trading from microgrid to grid is also a challenge. This is where decentralized energy systems and blockchain energy trading can revolutionize how energy is traded and used globally [2].

Microgrids are the solution to address the rapidly increasing demand while incorporating renewable energy sources. In decentralized systems, various energy sources such as solar photovoltaic (PV), wind turbines, biogas plants, hydropower systems, and energy storage systems like batteries, supercapacitors, and fuel cells are used. Energy generated by these systems is utilized locally, which reduces dependency on the centralized system. The microgrid can be operated on a grid or standalone. Standalone microgrids are important when the power supply to a place is very costly or when energy resilience is critical, such as natural disasters or grid failure. The microgrid can make the island from a utility grid when the central grid is down. The self-sufficiency in microgrids provides energy security to ensure uninterrupted supply during emergencies [3].

DOI: 10.1201/9781003545620-23

Advanced control systems are used in microgrids to allow real-time monitoring and optimization of energy flow and sufficient energy generation to match electrical load demand patterns. When renewable energy sources are insufficient to supply the load demand, energy storage devices must provide the energy needed to meet this deficiency [4]. Conversely, when energy production exceeds local consumption, grid-connected microgrids can feed excess power back into the larger grid, contributing to overall grid stability and energy efficiency.

To fully understand the technical aspects of microgrids, it is important to break down their core components.

18.1.1 Energy Generation Units

These include renewable sources like solar panels, wind turbines, and sometimes, backup generators that run on fossil fuels or biofuels for reliability. Including diverse energy sources within a microgrid allows for continuous power generation, even when renewable sources experience downtime.

18.1.2 Energy Storage Systems

Batteries are the most common form of energy storage in microgrids, allowing excess energy generated during peak production times (e.g., during the day for solar power) to be stored and used later when generation is low or demand is high. Technologies like lithium–ion batteries and emerging energy storage systems (e.g., solid-state and flow batteries) enhance the capacity and durability of energy storage solutions.

18.1.3 Power Distribution and Management

The energy generated and stored in a microgrid must be distributed effectively. This is accomplished through a sophisticated control system that monitors energy flows, ensures reliability, and balances supply and demand. Power electronic converters are utilized for AC–DC conversion in PV and batteries for AC loads. Figure 18.1 shows the energy trading in microgrid based on blockchain technology.

Utilizing peer-to-peer (P2P)-based energy trading in microgrids is revolutionary. Traditional consumers purchase power from the utility grid, which gets power from fossil fuel-based power-generating power plants [4]. Consumers who produce excess energy and sell it back to the grid are prosumers, as they consume and make the power. Decentralized energy trading can lead to a more dynamic and efficient market. It will lead to the optimization of local energy and promote renewable energy-based generation, which is environmentally friendly.

Blockchain leads to a secure, transparent, and decentralized ledger; transactions in energy trading between prosumers can be done without any intermediary [5]. Smart contracts are self-executing contracts in which terms are written directly into the code. These automatically handle the trading of energy when certain predefined conditions arise. Smart contracts offer seamless and low-cost transactions between participants, reducing dependency on traditional utility grid operators [6].

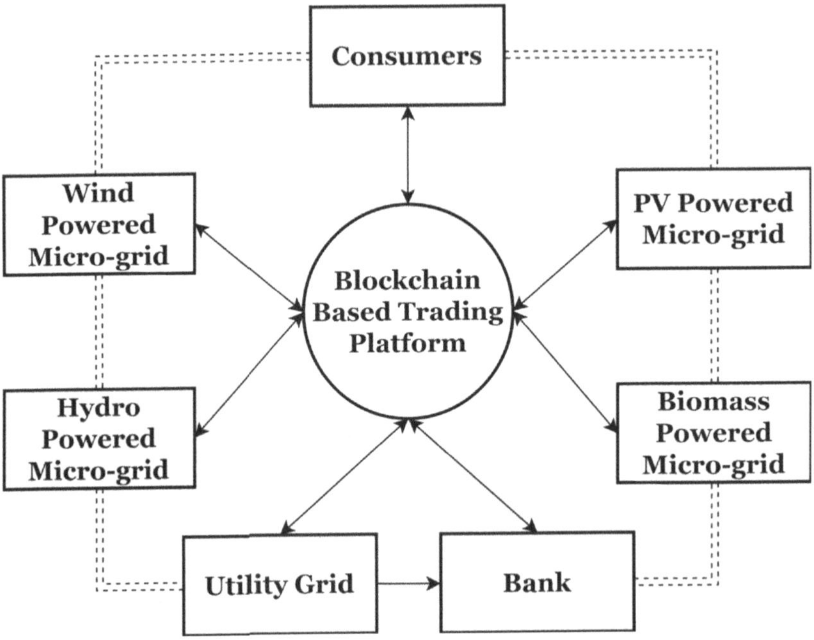

FIGURE 18.1 Blockchain-based energy trading in microgrids.

One of the good examples of P2P energy trading in a microgrid is Brooklyn Microgrid, New York. This project has attracted international interest because of the innovative system that is being used to allow local people to buy excess solar energy from each other using a blockchain. People who have installed PV systems on their rooftops can sell excess electricity to their fellow citizens without the involvement of power utilities [7]. The Brooklyn Microgrid develops an organized multiplayer energy market that will give numerous communities throughout New York City access to their energy ingestion and help decrease their carbon footprints.

Other successful early implementations exist besides the Brooklyn Microgrid, including Power Ledger of Australia. These systems promote the generation and utilization of renewable energy, less centralization of power generation, and hence inferior inefficiency as well as expenditures of conventional power generation and transmission [8].

Nevertheless, numerous studies recognize that utility-scale integration of microgrids is not without problems. First, there are high fixed costs when installing renewable energy power generation equipment like solar/Heating, Ventilation, and Air Conditioning (HVAC) panels, windmills, and energy storage system equipment. However, these technologies have come down in the recent past; the initial costs of infrastructure, control systems, and periodic maintenance could still add up and pose a significant cost burden, often to developing regions or small communities.

Government subsidies, Public-Private Partnerships (PPPs), and new innovative concepts like energy as a service platforms have assisted in some of these financial costs. However, to achieve broad penetration of microgrids, further reductions in the price of microgrid technologies are necessary, alongside improvement in awareness about the available financing options [9].

Second, there are such barriers as geographical and environmental factors. Some areas could be more conducive to renewable energy production. For instance, areas close to the seashore are likely to record high wind speeds, hence the suitability of wind turbines. At the same time, the mountainous or strongly wooded significantly lack direct sunlight, making solar panels suitable. In these cases, feasibility studies must be conducted to determine more appropriate energy generation methods on the particular sites.

Furthermore, using microgrids connected to the conventional centralized grid can pose legal issues and policies. Utilities and grid operators are typically slow to utilize decentralized systems because they threaten their current approaches. There is an increasing need for regulatory changes to allow new approaches in energy trading to cater to microgrids and P2P trading platforms.

Another challenge is the technological factor: the complexity of microgrids' technological systems and the technical know-how needed to run and support microgrids. Microgrid operators must have sufficient technical background to monitor the immediate dispatch of power, recharging, and grid synchronization [10]. For example, coastal areas might experience high wind speeds, making wind turbines ideal, but mountainous or heavily forested areas may have limited sunlight exposure, hindering solar panel efficiency. In these cases, site-specific assessments must be conducted to determine the best energy generation options.

Integrating microgrids with the existing centralized grid can create regulatory and policy challenges. Utilities and grid operators are often reluctant to embrace decentralized energy systems because they disrupt traditional business models. Regulatory frameworks must be updated to encourage innovation and accommodate new energy trading models that enable microgrids and P2P trading platforms to thrive.

Microgrids have certain challenges, and though they degrade their current and near-future picture, the future is incredibly bright for these systems. As costs for renewable energy technologies reduce and issues related to climate change become more well-known, microgrids are likely to emerge as one of the primary infrastructures for the future power system. Microgrids, providing localized generation and consumption of power, are a sustainable and low-vulnerability option to traditional centralized systems. The microgrid offers added advantages, including reduced greenhouse emissions, minimal transmission losses, and enhanced air quality, meeting the global objectives of combating climate change. Moreover, microgrids could be a source of new investment opportunities and fresh revenue generation for rural or other areas that experience energy deficits, thereby stimulating local energy generation and commerce.

The remaining chapter is organized as follows: Section 18.2 describes P2P energy trading, Section 18.3 highlights the working methodology of blockchain technology, Section 18.4 explores advancing sustainability through cryptocurrency-enabled energy trading, and Section 18.5 concludes the chapter.

18.2 PEER-TO-PEER ENERGY TRADING: BLOCKCHAIN AND CRYPTOCURRENCY

18.2.1 Blockchain and Cryptocurrency in Energy

Blockchain is another technology that has gained prominence with the popular use of coins such as bitcoins, and it has inherent characteristics that best fit into energy markets. A blockchain is a distributed database framework that establishes secure and transparent digital transactions without intermediaries [11]. It is also important in energy markets, where intermediaries like utilities or grid operators control the flow of electricity and money.

In a decentralized energy system, blockchain can provide the platform to perform energy trading through P2P where prosumers, which means the producer and consumer, can easily trade excess electricity with consumers [12]. Smart contracts, with programmed and self-implementing terms, accomplish these transactions to make energy transferred and paid instantaneously. Lastly, energy trading is protected and transparent due to the decentralized nature of blockchain [13]. Transaction data's immutability is secured from manipulation by cryptographic hashing and consensus protocols. However, through the minimization of intermediaries, blockchain has been able to minimize transaction discounts cost and optimizes the operational aspects of energy trading.

18.2.2 Role of Smart Contracts in Energy Markets

One of the most important elements of blockchain technology in energy trading is smart contracts. These smart contracts are calculated or agreed to be performed automatically once specific pre-stipulated conditions are met. For instance, a smart contract can sell extra electricity produced by the prosumer's solar panel and store it in a battery once it is full. Likewise, for instance, a consumer can sell electricity to them using the smart contract when the prices drops below a certain level [14]. Smart contracts remove the need for human intervention in energy trading, making transactions more efficient and less prone to error. In addition, because smart contracts are self-executing, they reduce the potential for disputes between parties, as the blockchain automatically enforces the contract terms. This automation can dramatically reduce the operational costs associated with energy trading [15]. Table 18.1 summarizes the key aspects of energy trading in microgrids and the role of blockchain technology, highlighting its benefits (e.g., decentralization, cost reduction) and challenges (e.g., scalability, regulatory barriers).

18.2.3 Energy Tokenization and the Emergence of Energy Cryptocurrencies

Tokenization is one of the important parts of blockchain regarding energy trading. It converts the physical energy (such as Kilowatt-hours) into digital tokens, which can be used for trading in blockchain. These tokens represent a quantity of electricity and can be exchanged by users in the decentralized market. It enables the creation of energy-specific cryptocurrency, which can be used for energy trading [16].

TABLE 18.1
Key Aspects of Energy Trading in Microgrids and the Role of Blockchain Technology

Aspect	Description	Applications	Limitations	Uniqueness
Decentralization	Microgrids operate locally, independently, or connected to the main grid.	Localized energy generation and consumption.	Coordination with the central grid can be complex.	Reduces dependence on central utilities.
Localized energy generation	Uses local renewable sources like solar and wind.	Lower transmission losses and enhanced energy autonomy.	Variability in renewable energy production	Promotes sustainable energy practices
Flexibility and resilience	Adapts to changes in energy needs and conditions and provides backup power.	Enhances grid reliability and rapid response	Requires robust infrastructure and management.	Improves grid stability and response to outages
Peer-to-peer trading	Direct transactions between participants without intermediaries.	Personalized and efficient energy transactions.	Regulatory and technical challenges in scaling.	Empower local energy markets and prosumers.
Secure transactions (blockchain)	Blockchain provides a decentralized, tamper-proof ledger	Ensures security and transparency in trading.	Potential for scalability and energy consumption issues.	Enhances trust and reduces fraud.
Smart contracts (blockchain)	Automated, self-executing contracts with terms written into code.	Automates and enforces trading agreements.	Complexity in designing and managing contracts.	Reduces administrative overhead and intermediary costs
Transparency and trust (blockchain)	Immutable records enhance transparency and verifiability.	Builds trust among participants and ensures fairness.	Requires comprehensive data management.	Provides an auditable and verifiable trading environment
Efficient settlement (blockchain)	Real-time transaction settlement reduces delays.	Improves cash flow and transaction efficiency.	Dependence on blockchain infrastructure stability.	Accelerates financial transactions and liquidity.
Cost reduction (blockchain)	Lower transaction fees and operational costs by eliminating intermediaries.	Reduces overall costs for energy trading.	Initial implementation and integration costs	It makes energy trading more economically viable.
Enhanced scalability (blockchain)	Blockchain supports scalable trading systems as the network grows.	Handles increased transaction volumes	Scaling can lead to increased energy consumption	Adapts to expanding microgrid networks

Power Ledger and major innovations enable users to buy and sell energy tokens directly without using traditional exchange techniques. These energy tokens can also be stored or transferred globally, allowing users to trade energy across borders seamlessly [17].

Moreover, tokenization provides a new investment opportunity for individuals and businesses. By purchasing energy tokens, investors can effectively fund renewable energy projects and benefit from the future sale of energy generated by those projects. This promotes the expansion of renewable energy and democratizes access to energy investments, allowing every day consumers to participate in energy markets [18].

18.2.4 Currency Mining and Energy Consumption

While blockchain technology offers numerous benefits for energy markets, it is important to acknowledge some of the challenges, particularly the energy-intensive nature of crypto currency mining. Mining is the process by which new transactions are verified and added to the blockchain. In networks like Bitcoin, mining involves solving complex mathematical problems, a process known as "proof-of-work," which requires significant computational power and, consequently, a substantial amount of electricity.

The high energy consumption of proof-of-work mining has raised concerns about the environmental impact of blockchain technology [19]. Take the example of Bitcoin, which alone consumes more energy over the years than some small nations, leading to the criticism that cryptocurrency mining contributes to the global energy demand instead of reducing it.

Since the blockchain is a leading technology and is helping in exploring more energy-efficient consensus mechanisms to address issues, Proof-of-Stake is one of the mechanisms that is significantly reducing the energy consumption by the user to "stake" a portion of their cryptocurrency holdings to confirm the transactions, instead of relying on energy-intensive consumption [18]. Ethereum's second-largest blockchain platform has recently transitioned to a proof-of-stake model by reducing energy consumption by more than 99%.

Furthermore, some blockchain projects are even exploring the idea of "green mining," using renewable energy sources to run their mining operations. These projects aim to reduce the environmental impact of cryptocurrency mining by utilizing solar, wind, or hydro-powered sources in combination with mining infrastructure to combine blockchain sustainability with wider sustainability objectives, security, and privacy in blockchain-based energy markets [19].

One of the essential benefits of blockchain technology in energy markets is that it affords elevated security—existing energy systems store transaction records in centralized databases and servers, which are susceptible to hacks and breaches. However, the distributed ledger of blockchain decentralizes the storage of transaction data across different nodes, which eliminates any single point of failure risk. For this reason, energy markets based on blockchain are, by nature, more secure.

In addition, blockchain employs state-of-the-art cryptography to secure data, making all transactions encrypted and human-verifiable. This maintains the integrity of the energy trading process and helps users improve their privacy. In legacy energy markets, customer data are usually centralized, and many players have access to it, opening the door to possible unauthorized data access. Blockchain provides a method where users can have more ownership and control of their data: only those with the correct cryptographic keys can access the information.

18.2.5 Role in Driving Energy Market Transparency

Another important aspect of blockchain in energy markets is transparency. Information regarding electricity generation, distribution, and pricing exists in silos and is controlled by centralized entities in traditional energy systems; thus, consumers find it difficult to access accurate information in real-time. So, this opacity triggers price manipulation and price dysfunctions and creates a trust gap in the energy market. By its nature, blockchain is decentralized, meaning all the participants have the same information simultaneously. This transparency allows consumers to make informed decisions regarding energy consumption and purchases [20]. For instance, consumers can observe the origin of the electricity they consume, whether from renewable or fossil energy sources, and the price of electricity in real-time.

18.3 WORKING WITH BLOCKCHAIN IN MICROGRID ENERGY TRADING

18.3.1 How Blockchain Works in Trading

Blockchain technology profoundly affects the operation of energy markets. In its most basic form, blockchain is a decentralized ledger that records transactions openly and securely across multiple computers [21]. Applying this to energy trading, blockchain facilitates the direct exchange of electricity between prosumers and consumers, removing intermediaries such as utility companies. Every transaction in a blockchain energy market is documented on a distributed ledger, which is open to all parties. This transparency makes it possible for all parties to trust that transactions are done correctly and protected by cryptographic hashing, making them unchangeable and tamper-proof. Smart contracts are an important part of energy transaction automation on blockchain platforms. These contracts are self-executing and get executed automatically when the predefined conditions are met [22]. For instance, if a prosumer has excess electricity and another consumer needs energy, a smart contract can facilitate the transaction automatically, transferring the energy and payment simultaneously.

18.3.2 Peer-to-Peer Trading Models

P2P energy trading is an innovative model in which individuals exchange electricity among themselves by buying and selling directly, bypassing the traditional utility companies. Several blockchain-based platforms have pioneered the idea of P2P energy trading, such as LO3 Energy and Power Ledger [23]. These allow prosumers to monetize their surplus energy through a blockchain-based marketplace, whereby they sell directly to nearby consumers.

Several potential benefits can be realized by P2P energy trading as opposed to the traditional energy markets, such as low transaction costs, high efficiency, and transparency improvements. It provides for more locally based energy production and consumption; thus, transmission losses are reduced while the grid's resilience increases. Table 18.2 provides a comparative analysis of blockchain-based technologies in energy trading, showcasing their applications (e.g., Power Ledger for P2P trading), limitations (e.g., regulatory barriers), and unique contributions to sustainability (e.g., smart contract automation).

18.4 ADVANCING SUSTAINABILITY THROUGH CRYPTOCURRENCY-ENABLED ENERGY TRADING

Cryptocurrency-enabled trading, powered by blockchain, big data, and artificial intelligence, represents a revolutionary approach to sustainability by facilitating the trading of surplus renewable energy in their preferred between customers—organizations that produce and use energy this alternative, such as solar and wind. Renewable energy adoption encourages and reduces reliance on fossil fuels and increases CO_2 emissions. Real-time energy trading improves the integration and quality of renewable energy, reduces energy waste, and contributes to grid efficiency. Case studies from India and Africa highlight the impact of blockchain-based microgrids in providing electricity, reducing carbon footprints, and generating economic growth in underserved areas. Furthermore, strategies of green mining and decentralized blockchain systems balance the production and consumption of energy and encourage the use of energy so that our environment and surroundings reinforce the role of cryptocurrency in sustainability. Table 18.3 describes the environmental impacts of cryptocurrency-enabled energy trading.

18.5 CONCLUSION

In this chapter, we have reviewed vast literature on the use of blockchain in microgrid decentralized energy trading. The study investigated how blockchain allows for P2P energy trading, improving the transparency, efficiency, and automation of transactions with smart contracts. This emphasized blockchain in renewable energy integration, providing grid flexibility and distributed energy autonomy. It also explored the advantages of time-of-use pricing and integration of variable renewable energy resources at the microgrid level.

It also points out the major issues, such as the scalability of blockchain networks, the energy consumption of traditional consensus mechanisms, and the regulatory complexities. It does so by critically examining these topics, emphasizing the need for more energy-efficient blockchain models, and inviting regulatory frameworks that will allow for decentralized trading. The work establishes a basis for future studies to investigate the technology of artificial intelligence and Internet of Things used in conjunction with blockchain to optimize the microgrid system and to evolve sustainability and the energy market.

TABLE 18.2
Blockchain-Based Technologies in Energy Trading and Sustainability [22, 24–28]

Technology Name	Application	Limitation	Unique Aspect
Power Ledger	The P2P energy trading platform for microgrids allows prosumers to trade excess renewable energy.	Regulatory barriers in some regions and scalability concerns due to blockchain energy usage.	Smart contract-based automated trading; transparent energy transactions.
Brooklyn Microgrid	Community-based microgrid using blockchain for local solar energy trading between neighbours.	It is limited to small-scale communities; regulatory challenges in scaling.	Local energy independence promotes the adoption of renewable energy sources.
Grid+	Decentralized and Autonomous (DeA) centralized energy market enables real-time energy pricing and trading for microgrid participants.	Relatively high operational costs for blockchain networks.	Real-time energy pricing; efficient energy use within microgrids.
WePower	A renewable energy trading platform connecting green energy producers with consumers uses blockchain.	Integrating with traditional grids is complex; adoption could be faster in some regions.	Tokenization of energy increases liquidity and accessibility of renewable energy.
SunContract	The P2P energy marketplace enables users to buy and sell renewable energy using blockchain.	Dependent on national regulatory frameworks and policies.	Direct connection between renewable producers and consumers; no middlemen.
Electrify	Facilitates energy trading between distributed energy resources within microgrids.	Limited pilot programs need broader integration into national grids.	Enables competition in energy pricing through blockchain; scalable for multiple energy sources.
TransActive Grid	Blockchain-based energy platform focusing on localized energy sharing in microgrids.	Needs advanced infrastructure like smart meters for full functionality.	Real-time energy distribution and trading; community-driven sustainability model.
Energy Web Chain	Open-source blockchain platform to accelerate low-carbon energy transition and decentralized trading.	High computational power requirements for blockchain operations.	Designed specifically for the energy sector, it focuses on regulatory compliance and decarbonization.
Share&Charge	Blockchain technology enables P2P trading of energy for electric vehicle charging within microgrids.	Limited to electric vehicle-related infrastructure; still emerging.	Facilitates the integration of electric vehicles with renewable microgrids.
LO3 Energy	Develops blockchain-based platforms for decentralized energy exchanges in community microgrids.	Early-stage development; challenges with scalability across large grids.	Community-driven energy trading; decentralized control over energy sources.

TABLE 18.3
Environmental Impacts of Cryptocurrency-Enabled Energy Trading

Aspect	Description	Impact on Sustainability
Carbon emissions reduction	Facilitates the use of renewable energy, reducing reliance on fossil fuels.	Decreases greenhouse gas emissions and carbon footprint.
Renewable energy integration	Optimizes the use of renewable energy sources through real-time trading.	Enhances the efficiency of renewable energy use.
Decentralization of energy production	Encourages localized energy production and consumption.	Reduces the need for centralized fossil-fuel-based generation.
Energy access (Africa)	Blockchain-based systems in Africa support renewable energy adoption and local trading.	Contributes to local economic development and sustainability.
Green mining	Utilizes renewable energy for crypto currency mining to reduce environmental impact.	Lowers the carbon footprint of mining operations.

REFERENCES

1. Y.-C. Tsao and V.-V. Thanh, "Toward blockchain-based renewable energy microgrid design considering default risk and demand uncertainty," *Renew. Energy*, vol. 163, pp. 870–881, 2021, doi: 10.1016/j.renene.2020.09.016.
2. K. Zhang, J. Yu, C. Lin, and J. Ning, "Blockchain-based access control for dynamic device management in microgrid," *Peer-to-Peer Netw. Appl.*, vol. 15, no. 3, pp. 1653–1668, 2022, doi: 10.1007/s12083-022-01316-5.
3. M. Bilal, A. A. Algethami, Imdadullah, and S. Hameed, "Review of computational intelligence approaches for microgrid energy management," *IEEE Access*, vol. 12, pp. 123294–123321, 2024, doi: 10.1109/access.2024.3440885.
4. S. Seven, G. Yao, A. Soran, A. Onen, and S. M. Muyeen, "Peer-to-peer energy trading in virtual power plant based on blockchain smart contracts," *IEEE Access*, vol. 8, pp. 175713–175726, 2020, doi: 10.1109/ACCESS.2020.3026180.
5. S. Rai, A. Mewada, M. A. Ansari, and D. Kumar, "Optimizing wireless sensor networks: Dynamic load balancing strategy with AOMDV routing," In *2024 3rd International Conference on Power Electronics and IoT Applications in Renewable Energy and Its Control (PARC)*, Mathura, India, 2024, pp. 350–355, doi: 10.1109/PARC59193.2024.10486598.
6. A. Waseem, M. Bilal, M. Danish, Imdadullah, and S. Hameed, "Revolutionizing Rural India: Blockchain-Powered Microgrid Management for Sustainable Development in India," In *2024 3rd International Conference on Power Electronics and IoT Applications in Renewable Energy and its Control (PARC)*, Feb. 2024, doi: 10.1109/parc59193.2024.10486555.
7. M. Sakib, S. Mustajab, K. Anwar, and M. Alam, "BiLSTM based approach for energy demand forecasting and synchronization among smart grids," In *2024 IEEE Third International Conference on Power Electronics, Intelligent Control and Energy Systems (ICPEICES)*, Delhi, India, 2024, pp. 969–974, doi: 10.1109/ICPEICES62430.2024.10719085.
8. R. S. Veena, G. Khambra, R. Sayal, N. Singh, et al., "An energy efficient IPv6 packet delivery outline for industrial IoT over G.9959 protocol based Wireless Sensor Network (WSN)," *SN Comput. Sci.*, vol. 5, p. 813, 2024, doi: 10.1007/s42979-024-03195-4.

9. Y. Wu, Y. Wu, H. Cimen, J. C. Vasquez, and J. M. Guerrero, "P2P energy trading: Blockchain-enabled P2P energy society with multi-scale flexibility services," *Energy Rep.*, vol. 8, pp. 3614–3628, 2022, doi: 10.1016/j.egyr.2022.02.074.
10. N. Singh, H. P. Singh, A. Mishra, A. Khare, M. Swarnkar, and S. K. Almas, "Blockchain cloud computing: Comparative study on DDoS, MITM and SQL injection attack," In *2024 IEEE International Conference on Big Data & Machine Learning (ICBDML)*, Bhopal, India, 2024, pp. 73–78, doi: 10.1109/ICBDML60909.2024.10577412.
11. X. Wang, F. Yao, and F. Wen, "Applications of blockchain technology in modern power systems: A brief survey," *Energies*, vol. 15, no. 13, p. 4516, 2022, doi: 10.3390/en15134516.
12. I. A. Umoren, S. S. A. Jaffary, M. Z. Shakir, K. Katzis, and H. Ahmadi, "Blockchain-based energy trading in electric-vehicle-enabled microgrids," *IEEE Consum. Electron. Mag.*, vol. 9, no. 6, pp. 66–71, 2020, doi: 10.1109/MCE.2020.2988904.
13. D. Han, C. Zhang, J. Ping, and Z. Yan, "Smart contract architecture for decentralized energy trading and management based on blockchains," *Energy*, vol. 199, p. 117417, 2020, doi: 10.1016/j.energy.2020.117417.
14. N. Singh, T. Chakrabarti, P. Chakrabarti, V. Panchenko, D. Budnikov, I. Yudaev, and V. Bolshev, "Analysis of heuristic optimization technique solutions for combined heat-power economic load dispatch," *Appl. Sci.*, vol 13, no. 18, p. 10380, 2023, doi: 10.3390/app131810380.
15. Z. Guan, X. Lu, W. Yang, L. Wu, N. Wang, and Z. Zhang, "Achieving efficient and Privacy-preserving energy trading based on blockchain and ABE in smart grid," *J. Parallel Distrib. Comput.*, vol. 147, pp. 34–45, 2021, doi: 10.1016/j.jpdc.2020.08.012.
16. O. Marin, T. Cioara, and I. Anghel, "Blockchain solution for buildings' multi-energy flexibility trading using multi-token standards," *Futur. Internet*, vol. 15, no. 5, p. 177, 2023, doi: 10.3390/fi15050177.
17. M. R. Islam, M. M. Rashid, M. A. Rahman, M. H. S. Bin Mohamad, and A. H. Bin Embong, "A comprehensive analysis of blockchain-based cryptocurrency mining impact on energy consumption," *Int. J. Adv. Comput. Sci. Appl.*, vol. 13, no. 4, pp. 590–598, 2022, doi: 10.14569/IJACSA.2022.0130469.
18. V. Sondhiya, K. Dwivedi, S. K. Almas, and N. Singh, "Energy management of nanogrid through flair of deep learning from IoT environments," In Nagendra Singh, Sitendra Tamrakar, Arvind Mewada, and Sanjeev Kumar Gupta (Eds.), *Artificial Intelligence Techniques in Power Systems Operations and Analysis*, Boca Raton, Florida: Auerbach Publications, Taylor& Francis, July 2023.
19. R. Alt and E. Wende, "Blockchain technology in energy markets – An interview with the European Energy Exchange," *Electron. M*ark., vol. 30, no. 2, pp. 325–330, 2020, doi: 10.1007/s12525-020-00423-6.
20. S. Shukla, D. Pandey, and N. Singh, "Grid interconnection of high step-up DC to AC converter for renewable source", In Gulshan Sharma, Pitshou N. Bokoro, and Sudeep Tanwar (Eds.), *Energy 4.0 Trends, Challenges, and Applications*, Boca Raton, Florida: CRC Press, 2024.
21. G. van Leeuwen, T. AlSkaif, M. Gibescu, and W. van Sark, "An integrated blockchain-based energy management platform with bilateral trading for microgrid communities," *Appl. Energy*, vol. 263, p. 114613, 2020, doi: 10.1016/j.apenergy.2020.114613.
22. N. Singh, A. Mishra, M. Gupta, K. Dwivedi, and M. Swarnkar, "Implementation of microgrid for rural electrification to cater on side load demand," In *2023 3rd International Conference on Energy, Power and Electrical Engineering (EPEE)*, Wuhan, China, 2023, pp. 1058–1061, doi: 10.1109/EPEE59859.2023.10351810.
23. H. T. Doan, J. Cho, and D. Kim, "Peer-to-peer energy trading in smart grid through blockchain: A double auction-based game theoretic approach," *IEEE Access*, vol. 9, 2021, doi: 10.1109/ACCESS.2021.3068730.

24. E. Mengelkamp, J. Gärttner, K. Rock, S. Kessler, L. Orsini, and C. Weinhardt, "Designing microgrid energy markets A case study: The Brooklyn Microgrid," *Appl. Energy*, vol. 210, pp. 870–880, 2018, doi: 10.1016/j.apenergy.2017.06.054.
25. A. Boumaiza, "A blockchain-based scalability solution with microgrids peer-to-peer trade," *Energies*, vol. 17, no. 4, p. 915, 2024, doi: 10.3390/en17040915.
26. Y. Mezquita, A. B. Gil-González, A. del Rey, J. Prieto, and J. M. Corchado, "Towards a blockchain-based peer-to-peer energy marketplace," *Energies*, vol. 15, no. 9, p. 3046, 2022, doi: 10.3390/en15093046.
27. M. Shahidehpour, M. Yan, P. Shikhar, S. Bahramirad, and A. Paaso, "Blockchain for peer-to-peer transactive energy trading in networked microgrids: providing an effective and decentralized strategy," *IEEE Electrif. Mag.*, vol. 8, no. 4, pp. 80–90, 2020, doi: 10.1109/MELE.2020.3026444.
28. N. Singh, S. K. Almas, R. Tirole, A. Jain, S. Shukla, and Y. Kumar, "Analysis of optimum cost and size of the hybrid power generation system using optimization technique," In *12th IEEE International Conference CSNT-2023 held at Technocrats Institute of Technology (Excellence)*, Bhopal, Madhya Pradesh, April 8 and 9, 2023. https://ieeexplore.ieee.org/document/10134720.

19 Blockchain for Social Impact
Shaping Future Management Practices

Richa Verma, Utkarsh Kumar Verma, Kalpana Rawat, Richa Srivastava, Arvind Mewada, Nagendra Singh, and Mohd. Aquib Ansari

19.1 INTRODUCTION

Blockchain technology, known for its decentralised digital ledger, has brought significant changes to various industries since the beginning of the Internet era. It offers a secure method for recording and verifying transactions across a network without third-party intermediaries, enhancing transparency and security [1]. Using cryptographic principles, blockchain creates a verified transaction history distributed across multiple network nodes. Recently, this technology has seen increasing application in social impact projects, prompting a reassessment of traditional governance, finance, and corporate social responsibility practices.

Blockchain is a distributed ledger technology that works on a peer-to-peer network where people can record and verify information safely and securely [2]. Among these features, decentralisation and immutability are the outstanding features. Decentralisation is defined as no central authority body giving control to a network of nodes. It prevents any entity from changing data and assures participant trust [3]. Over these raw themes, blockchain promotes transparency because each peer on the network is equipped with the same latest updated information at a real-time level. This openness can facilitate creating trust in systems that used to involve abstruse processes, which left avenues open for inefficiency or even some kind of abuse in the past [4].

19.1.1 Understanding Social Impact

Social impact can be described as the relative impact of any innovation or venture on society and its institutions- the relative effects on social welfare, equality, or justice. Technology's social impact looks into how blockchain technology results in positive social outcomes that manifest from greater openness, more decentralised control, and enhanced access to necessary services [5]. Furthermore, it allows everyone

DOI: 10.1201/9781003545620-24

access to full-fledged financial services in a traditionally bankless economic system by ensuring all can transfer money securely from peer to peer [6]. It can also raise the transparency and accessibility of services in sectors like health and education. For instance, blockchain technology used in healthcare can hasten patient data management; thereby, patients can easily manage their medical records while helping healthcare providers deliver better care [7]. Similarly, using blockchain to validate academic credentials in education can make it easier to share and authenticate qualifications across global networks.

19.1.2 Blockchain Meets Management Practice

Blockchain implementation into management practice has radical implications. Traditionally, the management system tended to be hierarchical, with commands on information flow and decision-making given by centralised authorities. Blockchain re-exposes all these traditional models with a decentralised peer-to-peer network for delegating control [2]. This decentralisation allows for more democratic management frameworks, enabling stakeholders to better view the goings-on in the decision-making processes and operations [1]. Within corporate settings, blockchain technology can improve the transparency of supply chains, strengthen CSR efforts, and ensure accountability across business operations. For example, blockchain-based supply chains will be able to track products from the starting points to the end, thus ensuring ethical sourcing practices and reducing inefficiencies in distribution [8]. Other than the above, blockchain provides transparency regarding CSR activities. The reason is that the activity records are maintained to check for any company initiatives that enhance the stakeholders' trust [9].

19.2 OBJECTIVES AND OUTLINE

The chapter highlights the following objectives:

- To explain the blockchain, elucidate its fundamental principles, and outline its most crucial issues.
- To discuss how blockchain can make a social impact in the governance, healthcare, finance, and education sectors.
- To analyse the transformative potential of blockchain in reframing the management attitude towards more openness, decentralisation, and inclusiveness in decision-making systems.
- To analyse issues and ethical dilemmas around integrating the technology of blockchain.

19.3 BLOCKCHAIN'S SOCIAL IMPACT

A **Governance Transformation:** Blockchain technology has significantly changed financial systems and, thus, introduced new governance approaches towards empowering individuals and communities through decentralised power structures. By establishing a distributed control model and institutionalising transparency, blockchain enables a structure for increased accountability and equity. The section discusses how blockchain's distributed nature transforms traditional governance structures and pushes social empowerment through inclusion, trust, fraud prevention, and access to essential services.

19.3.1 Decentralisation and Its Role in Social Empowerment

Decentralisation is the most transformative aspect of blockchain technology. Traditionally, centralised institutions have held power and jurisdiction, often leading to inefficiencies, corruption, and exclusion [10].

19.3.2 Recasting Tradition Governance Mechanisms

Decentralisation inherently changes governance mechanisms as no intermediaries or central authorities are required to make decisions. According to [1], because blockchain is inherently a network that makes decisions collectively, no one will be allowed to throttle the control mechanism in the decision-making process. From a top-down governance model, through a peer-to-peer network model, people and their groups are vested with much broader influence in the decisions that most impact them.

19.3.3 Inclusivity in Decision-Making

Blockchain technology allows all participants equal decision-making shares. It removes middlemen and offers unfettered access that other groups would not have had because of financial or bureaucratic constraints, according to [5].

19.3.4 Enhancing Transparency and Accountability

Blockchain technology improves transparency and accountability in many sectors. It does this by providing immutable and verifiable records of transactions and decisions. Blockchain is distributed to a network where every transaction is chronologically recorded and available to all. This feature makes alteration nearly impossible without majority consensus. Therefore, blockchain technology fosters greater accountability [4].

19.3.5 Blockchain as a Mechanism for Trust Building

Trust can be characterised as one of the most fundamental challenges of modern governance and management structures. Blockchain technology comes in handy by generating an immutable record of transactions that is visible to all network members. This makes it impossible to cheat or corrupt since any transaction gets an open and irreversible record. It is from such low-trust environments that public services consider; in them, opaque processes have often led to the erosion of public trust within governance structures. Fraud and corruption reduction through decentralised ledgers has been pivotal, with blockchain's decentralised ledger ensuring that transactions are transparent and verifiable [8]. All the above, any such fraudulent parties can detect and curb fraudulent actions since they must agree on validating a transaction before it hits the blockchain. For this reason, blockchain not only deters corrupt behaviours but also gives rise to transparent and accountable governance. This integrity is essential for the public to be confident in governmental and corporate systems [2].

19.3.6 Role of Blockchain in Alleviation of Social Inequality

Blockchain technology can significantly reduce social inequities by democratising access to financial systems and essential services. Traditional financial and governance models often exclude specific societal segments, affecting economic and social development. With its decentralised and all-inclusive methodology, blockchain offers new avenues for financial inclusion and egalitarian service access.

19.3.7 Financial Inclusion

Blockchain innovation financial inclusion is another area where blockchain plays a significant role. Since blockchain allows peers to transact with each other without banks directly, it also provides exposure to immense benefits offered to unbanked individuals, particularly in developing countries [11]. By affording a safe digital space to facilitate transactions, allowing those without access to traditional banking facilities to participate in economic activities becomes possible, leading to total financial inclusion [1].

19.3.8 Enabling Equal Access to Services in Underserved Communities

Blockchain technology promotes financial inclusion and enables access to essential services in underserved communities. For example, it can securely keep and verify personal identification, land titles, and learning documents. That, in turn, assists in one of the problems of acquiring said documentation in cases with difficulty in logistics or corruption problem issues [12]. Furthermore, blockchain technology enables patients to maintain and share health information securely; therefore, it increases the effectiveness of healthcare delivery while reducing health access inequities [7].

19.3.9 Blockchain and Management Practice Integration

Blockchain technology quickly rearranges technical and financial systems and fundamental business management processes. Its decentralised nature, transparency, accountability, and security allow one to rethink traditional management frameworks. Integrating blockchain into management practice transforms organisational governance and stakeholder management, contributing to corporate social responsibility and social impact investing.

19.3.10 Reconsidering Organisational Governance

Current forms of organisational governance are typified by structures where authority is centralised at the topmost level. Such frameworks are often riddled with problems of inefficiency, transparency problems, and decision-making processes that could be more convenient and take ages. However, the decentralised method of blockchain breaks the mould of conventional models where decisions may be dispersed over nodes or participants.

19.3.11 Distributed Decision-Making

In traditional corporate settings, decision-making is time-consuming, cumbersome, and involves approval from various hierarchical levels. The blockchain distributes decision-making among stakeholders instead. Through decentralised decisions, blockchain technology enables employees and partners, among other interested persons, to be actively involved in the governance process. The shift towards democratised decisions leads to inclusiveness and fairness in the organisational hierarchy, giving participants a say in crucial matters. Moreover, it decreases the probability of having a few individuals decide and decreases conflict of interest [13].

19.3.12 Stakeholder Management and the Blockchain

Stakeholder management typically involves the sensitive balancing of interests among numerous parties, such as employees, customers, investors, and regulators. Blockchain enhances stakeholder management by providing a transparent and immutable record of all engagement and transactions. That way, through transparency, all the stakeholders will have the same information, reducing the chance of miscommunication or tussles [8]. Furthermore, blockchain technology enhances stakeholder participation. In a decentralised system, stakeholders can voice their apprehensions about essential decisions by voting or presenting views on critical issues to ensure their interest is addressed. Shareholders may be able to participate at the policy-making level to determine company-level decisions or how a resource would be allocated in a blockchain network, which is then recorded on the blockchain [14].

This increases transparency and builds trust among stakeholders, who can verify that their contributions have been recognised. In addition, blockchain hastens stakeholder involvement through the automatic handling of the dividend disbursal process or profit-sharing processes. Automated discharge of dividends or profit-sharing functions can be capitalised on through smart contracts to automatically transfer funds to the stakeholders under certain conditions, thus facilitating accurate and timely payments [2]. With this automation in place, there is a reduction in administrative burdens because the processes relating to the management of stakeholders become effective and efficient [15].

19.3.13 Blockchain and CSR

In most CSR initiatives, issues with transparency and accountability arise as more stakeholders require extensive proof of a firm's impact on society and the environment.

Transparency is the foundation of successful CSR initiatives. Companies should prove the authenticity of their CSR activities instead of just treating them as promotional tools. Blockchain can be used in this regard. Blockchain can help a company trace and authenticate its entire products or service life cycle and thus authenticate CSR claims with verifiable data [8]. This transparency instils trust among consumers, investors, and other stakeholders by providing the means to verify the truth behind a company's claims visibly. In addition, the application of blockchain technology can also prevent such incidents as greenwashing, which involves falsely exaggerating or even fabricating environmental credentials by a company, thus sustaining the legitimacy of a CSR program, according to [9]. Blockchain can prove the money distribution in the CSR programs. For instance, if a company states that it will distribute a proportion of its profit to charity, blockchain can document the money flow and ensure that the money ends up in the target recipient's pockets [1]. The record in a blockchain places more legitimacy on the CSR activities, hence making companies responsible for their assertions [16].

19.3.14 Enhancing CSR Reporting Accountability

Investing has undergone significant transformation in recent years, with an increasing focus on creating social and environmental benefits alongside financial returns. Blockchain technology has become pivotal in ensuring transparency, accountability, and efficiency in managing social impact investments [17]. Investors can directly and securely support project-level or company-level initiatives by leveraging blockchain-based platforms. These platforms enable decentralised investments, often using cryptocurrencies or tokens to issue shares in projects or fund specific social initiatives [18]. For example, blockchain platforms can finance renewable energy projects by providing detailed, immutable records of each investment [19]. Blockchain integration addresses challenges often present in traditional investment frameworks, introducing unparalleled transparency and security. This transparency fosters trust among stakeholders and ensures that the impact of investments is accurately measured and verified [1]. Blockchain's ability to create an immutable history of transactions allows investors to track the impact of their contributions over time [18].

Furthermore, automating impact assessment and reporting processes reduces administrative costs, enhancing efficiency [15]. The success of such systems hinges on the quality of data input and storage within the blockchain, enabling data-driven decisions to refine strategies where necessary. Ultimately, the transparency and accountability facilitated by blockchain can ensure that social impact investments achieve their intended outcomes, delivering measurable and sustainable benefits [16].

19.4 BENEFITS OF BLOCKCHAIN TECHNOLOGY

In this part, we discuss the benefits of blockchain technology across industries.

19.4.1 Blockchain in SCM

Supply Chain Management (SCM) comprises complex networks with numerous mediators. Such a process has been chiefly inefficient, delayed, and non-transparent.

Blockchain tries overhauling supply chains by using live track records, guaranteed authentication, and building stakeholder trust. By providing unalterable registers, blockchain ensures equal access to information for any participant in the supply chain by reducing fraud, counterfeit, and unscrupulous sourcing.

19.4.2 Traceability and Ethical Sourcing

Blockchain has proven valid in offering traceability of products and ethically sourced products. This is of significant importance, and the challenge is based on consumer trust, making several ample food and retail companies embrace it. For instance, in collaboration with Walmart, IBM developed a blockchain-based food platform called the Food Trust. In 2018, the company made all suppliers in the leafy green category sign on to its blockchain network to strengthen food safety. The move helped the company trace contaminated products within seconds rather than days, reducing the risks associated with foodborne illnesses [20].

In addition, the collaboration brought about transparency as the customers can obtain truthful information concerning the history of the product; in this case, it is sourced from an ethical perspective. Nestlé also applied blockchain to its supply chain management. The company launched a blockchain system with Open SC that tracks from New Zealand farms and further down to the warehouses in the Middle East. This enables Nestlé to ensure that safety standards and ethical sourcing requirements are met at every point in the supply chain [21]. These examples illustrate how blockchain technology improves transparency in supply chains, ensuring responsible sourcing and meeting the ever-growing demand by consumers for ethically sourced products.

19.4.3 Cost Savings and Supply Chain Efficiency

Blockchain technology benefits business sectors as it becomes more transparent and efficient. For example, Maersk and IBM collaborated to develop Trade Lens, a real-time container tracking system. This enables players in the supply chain to collaborate. With automated documentation processes and digitisation of paperwork, Maerske was able to reduce costs while simultaneously improving operational efficiency in shipping. De Beers applied technology with Tracr, too, in its diamonds' traceability so that they appear authentic and sourced ethically. This program provides transparency and functional improvements by tracing diamond origin in real-time. These illustrations explain how the future of blockchain innovation could reduce cost, expedite processes, and improve safety in supply chain procedures.

19.4.4 Blockchain Impact on Management of Security in Health Records

Health is characterised by the volume of information that should be managed securely and effectively. The most essential healthcare issue is the clear assurance about properly managing healthcare records. Blockchain technology allows healthcare patients to safely manage their health data while allowing authorised providers access to records. A representative example in the healthcare industry is the Medi Ledger

Project, an initiative undertaken by drug majors such as Pfizer and Genentech that uses blockchain technology to authenticate and track drugs so that spurious medications cannot find their way into the supply chain [22].

19.4.5 Blockchain Ensuring Better Security to Health Records and Patient Confidentiality

Blockchain alleviates these issues by providing a decentralised and secure method of managing health records, significantly limiting the chance of unauthorised access or data tampering. Estonia has used a blockchain-based system headed by the e-Health Foundation to secure health records for its citizens. This blockchain framework ensures that all interactions between patients and providers are recorded providently in an immutable ledger, ensuring full transparency and accountability [22]. Estonia's blockchain solution ensures that any particular health record is accessible only to authorised parties, significantly improving patient privacy without compromising transparency. The firm based in Estonia, specialising in blockchain technology, Guard Time, has developed the KSI Blockchain, which is used to protect medical records and the integrity of information in most healthcare systems. The blockchain product from Guard Time ensures that medical records are secure by preventing data breaches and ensuring information about patients is kept as secret.

19.4.6 Education: Blockchain-Based Credentials and Accessibility

Blockchain technology has much to say for the education sector, primarily verifying academic credentials and enhancing accessibility in education. This technology ensures integrity via a blockchain-based secure ledger of educational accomplishments, resisting tampering and allowing easy verification processes. This not only eases procedures intended to reduce fraud but also cuts down on the hassle quotient entailed by verifications by employers or institutions of education.

19.4.7 Verification of Educational Documents with Blockchain Technology

Verifying educational credentials is lengthy, costly, and, in most cases, inaccurate. Blockchain technologies will soon end this situation when academic institutions can verify the authenticity of digital diplomas and certificates on an irreversible ledger. MIT has pioneered this innovation by introducing its Digital Diploma Initiative, which allows graduates to obtain blockchain-backed digital diplomas. This technological advancement enables the recruiter to verify qualifications easily and guarantees the proper hiring of people with legal certificates. Sony Global Education ventured into blockchain by creating a platform that, among other things, would assist in protecting and transferring academic data. This ground-breaking platform enables students to share their official, authenticated academic achievements with relevant organisations and employers across the globe to ease the processing of admissions and recruitment [23].

19.4.8 Blockchain as the Answer to Expanding Access to Education in Developing Regions

Blockchain verifies the authenticity of documents and significantly widens educational opportunities, especially in far-flung places restricted from accessing regular education systems. Blockchain allows students to be part of international educational structures, and this is made possible through decentralised systems for checking students' records and authenticating attendance and other grades. This adoption is seen through African Broadcasting Union (ABU) using blockchain to track students' files and provide academic materials to students across African nations. Students can showcase their credentials to prospective employers and educational bodies through the blockchain-based platform accessible through ABU, thus equipping them with the resources necessary to gain a suitable advantage on global playing fields [24].

19.4.9 Design of Social Business Models through Blockchain

Blockchain technology helps social entrepreneurs design decentralised business models that empower communities and amplify the sense of transparency. Blockchain technology ensures efficient allocation of resources and transparent operations, thus creating trust and accountable stakeholders. For example, BanQu is a blockchain technology that allows poor people to build an economic identity based on recorded transactions. BanQu enables marginalised groups to enhance their livelihoods through access to financial services and global supply chains without being involved in corrupt practices in handling resources; instead, such practice is transparent [19]. This platform connects smallholder farmers and farmers with global markets, facilitating fair trade practices and economic opportunities. In addition, the blockchain-based-provenance platform Provenance supports social entrepreneurs by making transparent and traceable supply chain solutions available for producing socially beneficial products. Through Provenance, consumers can trace the product origin and the product while confirming positive impact support for the ethical businesses assured to be supported [25].

19.4.10 Crowdfunding and Impact Tracking via Decentralised Platforms

The social entrepreneur can access an investor community globally via decentralised crowdfunding, whereas full transparency in funds use can be guaranteed. Using such a website, donors can see where their money goes through a public, decentralised ledger; that way, all distributions of funds meet the goals of donors [6]. This is meant to create trust and transparency on the part of the donor, who gets to view the projects they are funding in real-time. Most importantly, blockchain technology helps track the social and environmental impact behind every undertaking with tremendous accuracy. This facilitates accurate data being provided to patrons and stakeholders, elevating credibility.

19.4.11 Challenges and Ethical Considerations in Blockchain Implementation

Blockchain technology has revolutionised many industries with decentralised, transparent, and secure systems. Besides its promising benefits, blockchain implementation faces significant challenges and ethical issues. These include privacy concerns, the challenge of legal and regulatory problems, ecological impacts, and the ethical use of blockchain in vulnerable societies.

19.4.12 Balancing Transparency with Data Privacy Issue

Transparency is one of its core strengths; all transactions are visible on the ledger. However, this quality sometimes conflicts with privacy requirements, especially with sensitive personal or financial data. In cases like public blockchains, entire transaction histories are open to everyone, potentially revealing personal information inadvertently. The dilemma here causes a significant issue, especially in health care and finance industries where privacy is an issue.

19.4.13 Solutions for Protecting Sensitive Information on Blockchains

Using the most advanced encryption and privacy-enhancing technologies to protect sensitive information is crucial. Zero-knowledge proofs and secure multi-party computation are emerging solutions. Zero-knowledge proofs enable a party to verify that a statement is true without revealing the underlying data, thereby preventing the leakage of secret information.

19.4.14 Discovery of Sustainable Alternatives

The effort to minimise the environmental footprint of blockchain technology led to the discovery of energy-efficient variants for consensus mechanisms and hybrid approaches. For instance, Ethereum is the blockchain network that mostly migrates from PoW to PoS, with the Ethereum 2.0 upgrade reducing its energy consumption by up to 99.95% [8]. Other environment-friendly options involve Proof of Authority and delegated Proof of Stake, adopted by platforms like VeChain and EOS, to offer reduced resource demand.

19.4.15 Ensuring Ethical Use in Vulnerable Communities

Blockchain solutions must be guided by responsible technology that will help assist vulnerable groups, such as refugees, unbanked accounts, etc. However, such endeavours must address surveillance risks, data exploitation, and misapplication by governmental or other forces. Ethical issues align with members of vulnerable communities maintaining control over their data, while blockchain applications must not enhance current inequalities.

19.5 FUTURE DIRECTIONS

Blockchain technology is an evolving innovation with substantial potential to transform sectors by advancing social impact, governance, and management practices. Its integration with cutting-edge technologies such as artificial intelligence and Internet of Things (IoT) enhances its application, enabling automation, real-time data collection, and secure decision-making processes. For example, blockchain and IoT ensure supply chain transparency, particularly in agriculture and healthcare, by tracking products' origins and ensuring compliance with ethical standards [26]. Estonia's adoption of blockchain in e-governance exemplifies its capacity to simplify administration and ensure transparency. Decentralised autonomous organisations further foster participatory governance by enabling decentralised decision-making. In business, blockchain promotes sustainable and ethical practices by enhancing trust through immutable records. Platforms like Provenance demonstrate how blockchain can verify supply chain claims, reinforcing corporate social responsibility. Its functionality in automating operations through smart contracts boosts efficiency, while shared ledgers support collaboration, as seen in IBM's blockchain application in global supply chains. By aligning ethical standards with operational efficiency, blockchain technology has the potential to reshape governance and management practices, ensuring greater transparency and fostering sustainable growth.

19.6 CONCLUSION

Decentralised, transparent, and accountable, blockchain offers an excellent force for social impact. Through blockchain, real-time data monitoring and validation in support of ethical sourcing, the disarming of fraud, and the ability of businesses and organisations to take the sustainability game an extra step further in demonstrating it is possible. The immutability of records in blockchain lends credibility and assurance among various stakeholders towards implementing CSR initiatives and sustainability goals. More than this, blockchain will change governance and management practices when innovative solutions focus on absolute transparency, fight corruption, and empower the marginalised. Decentralised governance structures enable individuals and communities to participate in decision-making for inclusivity and accountability. Blockchain has the power to support sustainable business models and ethical behaviours; hence, it could be something that businesses and governments look to grapple with issues in society and the environment seriously. So, it can steer the pathway towards a fairer and more sustainable future in this way through its progress and adoption.

REFERENCES

1. Singh, N., Singh, H. P., Mishra, A., Khare, A., Swarnkar, M., & Almas, S. K. "Blockchain cloud computing: Comparative study on DDoS, MITM and SQL injection attack", In *2024 IEEE International Conference on Big Data & Machine Learning (ICBDML)*, Bhopal, India, pp. 73–78, 2024. https://doi.org/10.1109/ICBDML60909.2024.10577412.
2. Iansiti, M., & Lakhani, K. R. "The truth about blockchain", *Harvard Business Review*, 95(1), 118–127, 2017.

3. Singh, N., & Kumar, Y. "Security and privacy requirements in 5G healthcare", In *Block Chain for 5G Healthcare Applications.* IET Publication, pp. 1–30, Nov. 2021. https://doi.org/10.1049/PBHE035E_ch1.
4. Zheng, Z., Xie, S., Dai, H., Chen, X., & Wang, H. "An overview of blockchain technology: Architecture, consensus, and future trends", In *2017 IEEE International Congress on Big Data (BigData Congress)*, pp. 557–564, 2017. https://doi.org/10.1109/BigDataCongress.2017.85.
5. Veena, R. S., Khambra, G., Sayal, R., Singh, N., Dhandayuthapani, V. B., & Sar, S. K. "An energy efficient IPv6 packet delivery outline for industrial IoT over G.9959 protocol based Wireless Sensor Network (WSN)", *SN Computer Science,* 5, 813, 2024. https://doi.org/10.1007/s42979-024-03195-4.
6. Dewang, R. K., Yadav, M. P., Awasthi, S., Raj, O., Mewada, A., & Bawankule, K. L. "Data secure application: An application that allows developers to store user data securely using blockchain and IPFS", *Multimedia Tools and Applications*, 83(15), 45491–45517, 2024.
7. Azaria, A., Ekblaw, A., Vieira, T., & Lippman, A. "MedRec: Using blockchain for medical data access and permission management", In *2016 2nd International Conference on Open and Big Data (OBD)*, pp. 25–30, 2016. https://doi.org/10.1109/OBD.2016.11.
8. Saberi, S., Kouhizadeh, M., Sarkis, J., & Shen, L. "Blockchain technology and its relationships to sustainable supply chain management", *International Journal of Production Research*, 57(7), 2117–2135, 2019. https://doi.org/10.1080/00207543.2018.1533261.
9. Morkunas, V. J., Paschen, J., & Boon, E. "How blockchain technologies impact your business model", *Business Horizons*, 62(3), 295–306, 2019. https://doi.org/10.1016/j.bushor.2019.01.009.
10. Singh, N., Sasirekha, S. P., Dhakne, A., Sai Thrinath, B. V., Ramya, D., & Thiagarajan, R. "IOT enabled hybrid model with learning ability for E-health care systems", *Measurement: Sensors*, 24, 1–14, 2022. https://doi.org/10.1016/j.measen.2022.100567.
11. Zhu, H., & Zhou, Z. Z. "Analysis and outlook of applications of blockchain technology to equity crowdfunding in China", *Financial Innovation*, 2(1), 1–13, 2016. https://doi.org/10.1186/s40854-016-0044-7
12. Sharples, M., & Domingue, J. "The blockchain and kudos: A distributed system for educational record, reputation and reward", In *Adaptive and Adaptable Learning: 11th European Conference on Technology Enhanced Learning, EC-TEL 2016, Lyon, France, September 13–16, 2016, Proceedings 11*, pp. 490–496, 2016.
13. Beck, R., Avital, M., Rossi, M., & Thatcher, J. B. "Blockchain technology in business information systems research", *Business & Information Systems Engineering*, 59(6), 381–384, 2017. https://doi.org/10.1007/s12599-017-0505-1.
14. Goenkaa, N., Sharmab, A. K., Tiwaric, S., Singhd, N., Yadave, V., Prabhuf, S., & Chadaga, K. "A regularized volumetric ConvNet based Alzheimer detection using T1-weighted MRI images", *Cogent Engineering*, 11(1), 2314872, 2024. https://doi.org/10.1080/23311916.2024.2314872.
15. Nofer, M., Gomber, P., Hinz, O., & Schiereck, D. "Blockchain", *Business & Information Systems Engineering*, 59(3), 183–187, 2017. https://doi.org/10.1007/s12599-017-0467-3.
16. Di Vaio, A., & Varriale, L. "Blockchain technology in supply chain management for sustainable performance: Evidence from the airport industry", *International Journal of Information Management*, 52, 102014, 2020. https://doi.org/10.1016/j.ijinfomgt.2019.09.010.
17. Dutta, P., Choi, T.-M., Somani, S., & Butala, R. "Blockchain technology in supply chain operations: Applications, challenges, and research opportunities", *Transportation Research Part E: Logistics and Transportation Review*, 142, 102067, 2020. https://doi.org/10.1016/j.tre.2020.102067.

18. Karthikeyan, D., Amutha, B., Devi, K. M., Singh, N., & Kuruba, P. A. "Hybrid rater to quantify and measure the severity of infection and spread of infection in muskmelon", *Bulletin of Electrical Engineering and Informatics (BEEI)*, 13(3), 1667–1675, 2024. https://doi.org/10.11591/eei.v13i3.5432.
19. Kim, H. M., & Laskowski, M. "Toward an ontology-driven blockchain design for supply-chain provenance", *Intelligent Systems in Accounting, Finance, and Management*, 25(1), 18–27, 2018. https://doi.org/10.1002/isaf.1424.
20. Kamath, R. "Food traceability on blockchain: Walmart's pork and mango pilots with IBM", *The Journal of the British Blockchain Association*, 1(1), 3712, 2018. https://doi.org/10.31585/jbba-1-1-(10)2018.
21. Faiyaz, M. S., Srivastava, G., Singh, S., & Kumar, S. "Emerging trend in supply chain blockchain and Industry 4.0," In *Supply Chain Management*, CRC Press, pp. 93–118, 2024.
22. Mettler, M. "Blockchain technology in healthcare: The revolution starts here", In *2016 IEEE 18th International Conference on E-Health Networking, Applications and Services (Healthcom)*, Munich, Germany, pp. 1–3, 2016. https://doi.org/10.1109/HealthCom.2016.7749510.
23. Henderson, A., & Romeike, P. D. "Blockchain in education: Opportunities, applications, and threats", *European Journal of Open, Distance and E-learning*, 21(2), 73–81, 2018. https://doi.org/10.2478/eurodl-2018-0006.
24. Boukis, A. "Exploring the implications of blockchain technology for brand-consumer relationships: A future research agenda", *Journal of Product & Brand Management*, 28(3), 258–272, 2019. https://doi.org/10.1108/JPBM-03-2018-1782.
25. Ben-Sasson, E., Chiesa, A., Garman, C., Green, M., Miers, I., Tromer, E., & Virza, M. "Zerocash: Decentralised anonymous payments from Bitcoin", In *2014 IEEE Symposium on Security and Privacy*. Berkeley, CA, USA: IEEE, pp. 459–474, 2014.
26. Rupesh Kumar Dewang, Mahendra Pratap Yadav, Surbhit Awasthi, Om Raj, Arvind Mewada, Kamlakant Laxman Bawankule "Data secure application: an application that allows developers to store user data securely using blockchain and IPFS". *Multimed Tools Appl* 83, 45491–45517 (2024). https://doi.org/10.1007/s11042-023-17204-w

20 Planting the Seeds of Reform

The Effect of Blockchain on Seed Management

Sumit Kumar Sharma, Bhawna, Arvind Mewada, and Mohd. Aquib Ansari

20.1 INTRODUCTION

Food security is a critical pillar of any economy and it depends on a food supply chain that is connected to the management of farming resources. This means that the ineffective management of farming resources not only impacts yield but also impacts food security in any nation. Seeds are invaluable assets in farming; their quality and availability are key parameters for evaluating the farming process. Seed accessibility to farmers not only improves yield but also helps farmers to make informed decisions. Traditional supply chain methods hamper the operation of processes due to the involvement of external parties like middlemen; the addition of fake seeds in the supply chain is a reason for making this process unreliable and opaque.

In addition, malpractices of using researchers' work without their permission, which leads to intellectual property rights infringement cases, add delay and choices in the supply chain. These challenges open the door for the integration of emerging technology solutions in this space that improve the seed supply chain management and increase the farmer's confidence in the quality of the seeds in supply. Technology seems to be a transformative technology in space that has the ability to revolutionize various processes, including agriculture. Its immutable and decentralized nature offers a safe and transparent platform for verifying and recording transactions. By leveraging blockchain's resources, it is possible to tackle the challenges faced in established seed management systems, such as traceability, validity, and intellectual property rights [1]. The international industry is responsible for managing the supply of high-quality seeds that help in the production of crops in order to ensure food security; however, there have been common challenges in this industry since its inception: those are intellectual property, issues in the supply chain and quality, damages due to transportation, and malpractices in the availability of high-quality seeds in general [2,3].

Another factor that impacts the smooth delivery of the seeds is the complicated nature of the factors, such as changing the preferences of the farmers from quality seeds to high-producing seeds, regulations, and changes in climate that impact the

 DOI: 10.1201/9781003545620-25

soil and ultimately affect the seed supplies. These fast changes that impact seed supplies create a need for the implementation of new-age technology solutions in managing the supply of seeds that make them traceable, add transparency, and also make the actors in the supply chain accountable for their role in the chain [4,5]. As blockchain makes the process distributed via assigning a hash to every actor when an actor performs any action in the chain in order to record the transactions performed, as blockchain technology creates the decentralized digital ledger, it manages the transactions with accountability, traceability, and reliability over the responses, which makes this technology enhance the trust, efficiency, and traceability in processes such as supply chain management [6].

20.2 UNDERSTANDING THE SEED SUPPLY CHAIN

Various stakeholders are involved in the management of the complex process of supply chain management of seeds from manufacturers to farmers, including the following actors with the challenges particular actors are suffering from [7]. Seed producers are primarily institution-owned laboratories, whether private or public, that use various research and development techniques to produce high-quality seeds, sometimes also known as seed breeders. Challenges include competition with other firms and organizations, stealing of their intellectual property rights (IPR) [8,9], and high costs associated with the research in producing various new breeds of crop seeds based on the problems that arise or solving under any project. Certification institutions play a critical role whenever new seed developed by producers needs to be approved by a certifying government body that gives that certification for regular consumer market selling or distributing, and the problem-fighting certification authority is doping in samples submitted for testing, parallel balancing the high quality and standards of grading the seeds. Seeds are certified and then collected by wholesalers, who are responsible for distributing them to retailers or farmers [10].

Challenges faced by wholesalers include eliminating the middleman in the procurement of seeds from producers, which increases the cost of seeds at the source only and further impacts the margins of the wholesaler and retailers in the chain. Retailers are major distributors from which direct farmers or generic customers purchase the seeds for sowing in the crop fields. Inventory management and storing the seeds in the required storage type and cost in that required stage are the primary challenges these retailers are continuously fighting [11].

The farmer is an actor who is a major consumer of the seeds for harvesting desired products based on the location and climate of crop fields, frequent changes in climate, soil properties, and the price of the seeds in the supply chain system of seeds are the major issues farmers are suffering from. Another major challenge that impacts the complete supply management of the seed industry is the bookkeeping of entries of seeds [5], which includes data fragmentation, which means data related to the transaction of seeds is stored in various storages and even different formats that make it difficult to process collectively. Data reliability and precision over the data stored are other concerns that make the evaluation metric unreliable, as standard metrics in this case. Gap and traceability are not practically possible without a reliable source of entries at every point from where seeds pass through, which ultimately impacts the quality of the seeds.

Data security is a major issue that impacts the supply chain most, apart from the data recording challenges faced by the current supply chain management process. One solution for all current challenges, as mentioned above in seeds supply chain management [12,13], is blockchain technology. It provides various features for automatically recording each step in a distributed manner, ensuring that every actor in the supply chain is informed and able to track the status of each batch of seeds as it moves through the supply process. It tracks the distribution of crops in real time, helping government agencies manage demand and supply balance, control prices for major crops, and oversee the harvesting and procurement systems in the specified areas [14,15].

This chapter aims to explore the application of blockchain technology in seed management. It will investigate how blockchain can improve seed traceability, guaranteeing that seeds are traced from legitimate suppliers and sustain quality standards. The proposed study summarizes the perspective of blockchain to protect the IPRs of stakeholders such as seed breeders and producers. By analysing these aspects, the study provides insight into seed management, where blockchain technology plays a crucial role in adding reliability to supply chain management.

20.3 CHALLENGES IN SEED SUPPLY CHAIN MANAGEMENT

Seed supply management has a complex network of stakeholders, incorporating seed breeders, regulatory bodies, distributors, retailers, and farmers. Each stakeholder plays a vital role at different stages in confirming the authenticity, quality, and timely delivery of seeds [16,17]. However, the established seed supply chain encounters numerous challenges that can hamper its efficiency and effectiveness. Key challenges faced by stakeholders include the scarcity of traceability, which is one of the prime concerns. There is a significant need for robust traceability systems to track the source, journey, and quality of seeds throughout the supply chain [15].

Data fragmentation is another issue, where information about seed quality, production, and distribution is often scattered and fragmented across several systems. This fragmentation obstructs data analysis and decision-making. Quality issues also present a major concern, as confirming seed quality is essential. Contamination, counterfeit seeds, and low-quality seeds can drastically impact crop yields and farmer earnings. IPR safety is another critical challenge, as protecting IPRs associated with seed varieties is crucial for innovation and security. However, challenges in enforcing IPRs can lead to piracy and illegal seed production. The challenges mentioned above can have substantial negative impacts on the seed supply chain. Lack of traceability can hinder deliveries, increase costs, and erode trust among stakeholders [18,19]. Data fragmentation impedes data-driven decision-making and blocks the identification of trends and patterns. Quality issues can lead to crop failures, financial losses for farmers, and damage to the reputation of seed manufacturers [20]. Furthermore, inadequate IPR protection can suppress innovation and restrict investment in seed research and development.

20.4 BLOCKCHAIN AS A SOLUTION FOR SEED MANAGEMENT

Blockchain is an innovative technology that permits secure and evident recording of transactions among multiple computers. It acts as a distributed record, where data is collected in blocks that are linked simultaneously using cryptography. This produces a chain of cell blocks, hence the name "Blockchain." This technology is an immutable, decentralized, and transparent digital ledger that documents transactions across multiple computers. Every transaction is inserted into a block, linked to preceding blocks, creating a chain [16,21].

This chain of blocks is distributed throughout a network of computers, making it substantially secure and resistant to tampering. The aim for seed management through blockchain's key highlights consists of immutability, ensuring that once data is authenticated on a blockchain, it cannot be amended or deleted, thereby safeguarding the integrity of seed records. Transparency ensures that each transaction on the blockchain is noticeable to all participants, progressing accountability and transparency. Traceability in blockchain enables the tracking of seeds from production to utilization, reducing the risk of counterfeit or low-quality seeds and enhancing trust in the supply chain. The primary flow of seeds begins with the breeder as the first actor and extends to the farmer as the end consumer.

This process can be effectively monitored using several key methods. Contract administration involves maintaining all information linked to the licensee, roles, and the logging of actions. Payment administration ensures that every transaction is recorded and that stock books are updated accurately. Seed quality management focuses on tracking high-demand seeds, enabling efficient real-time supply chain management.

In addition, verification and IPR management ensure that all transactions are validated by actors on the blockchain. At the same time, IPRs are securely preserved for the author, who retains hash ownership within the chain. A classified blockchain network can be inaugurated to facilitate secure and efficient seed management [22]. This network can engage key stakeholders such as seed producers, retailers, wholesalers, certification authorities, and farmers. Smart contracts, or self-executing contracts with the terms of the agreement directly transcribed into code, can automate several processes, such as payment transactions, quality constraint checks, and IPR licensing.

Key benefits of blockchain for seed management include improved traceability, which is achieved through blockchain, facilitating the tracking of seed extraction from the source to the end-user. Each step of the seed journey, including production, processing, packing, and allocation, can be documented on the blockchain [23]. This enhances transparency and responsibility while reducing the risk of counterfeit or inferior seeds. Superior IPR protection is another advantage, as blockchain can better safeguard IPRs associated with seed varieties. By recording ownership details, licensing contracts, and royalty payments on the blockchain, it becomes significantly harder to infringe on IPRs.

In addition, the distributed and immutable nature of blockchain ensures the security and reliability of seed data. All transactions are recorded on the blockchain, making it difficult to manipulate or alter records [24]. This strengthens accountability among all stakeholders and fosters trust in the seed supply chain.

20.5 IMPLEMENTATION FRAMEWORK

The implementation of blockchain that is able to handle the primary problems associated with seed supply management requires the below-defined steps and execution logics, which might vary for each use case and environment. In order to implement blockchain, several steps are required for blockchain integration. The process begins with the identification of actors/stakeholders and their roles, including farmers, seed producers, regulatory bodies, distributors, retailers, and technology support providers. Next, the type of blockchain needs to be defined, and a private blockchain, suitable for stakeholders within a single organization, or a public blockchain, which accommodates a large number of stakeholders and ensures greater transparency, needs to be selected. Smart contracts are then set up, establishing the rules required for actions such as seed certification, financial transactions, and quality checks.

A compatibility check follows, ensuring that the system is designed in a way that does not affect the existing workflow while seamlessly connecting with the current system. Additional steps include designing the system architecture and defining all actors, actions, and data layers for both on-chain and off-chain data points. Key data points such as seed ID, variety, quality attributes, expiry date, source details, consumer details, and location should be defined, along with tracking standards like GS1 and tokenization, which may involve fungible or non-fungible tokens for each seed batch.

Smart contract functions are developed for various actions, such as registering, storing, certifying, and validating, along with the logic for triggering events that update the hash whenever executed. Infrastructure implementation involves deploying the network, adding/removing nodes with permissions, and integrating wallets for digital identity. Frontend development is required for user interactions, while the backend enables communication between the front end and the core functionalities of the application. Finally, the system undergoes testing, monitoring, and maintenance, with regular contract updates as needed.

The technical components of the blockchain-based system include smart contracts, which automate and enforce the rules for certification, payments, tracing, and compliance checks. A key component of this is the ERC-721 token, which serves as a template that developers must follow when creating the authentication mechanism for the blockchain system [25]. For example, a smart contract processes payment only for verified seed deliveries. Tokenization involves assigning unique codes to seed batches, enabling the tracing of these batches and ensuring security. Cryptographic methods are used to manage seed data integrity and security through techniques like hashing and digital signatures, which allow the safe movement of data within the environment.

Distributed storage ensures that all data points associated with seeds, such as certifications, Internet of Things (IoT) monitoring logs, and transaction records, are safely stored in a distributed system, protecting them from data loss due to failures. Other significant attributes of the system include the ledger, which must be immutable to prevent any tampering or changes to the records. Access control is implemented to assign permissions based on roles, such as regulators, who are the only ones authorized to assign certificate hash keys. The integration environment may utilize platforms like Oracle or SAP to fetch the data required for the process.

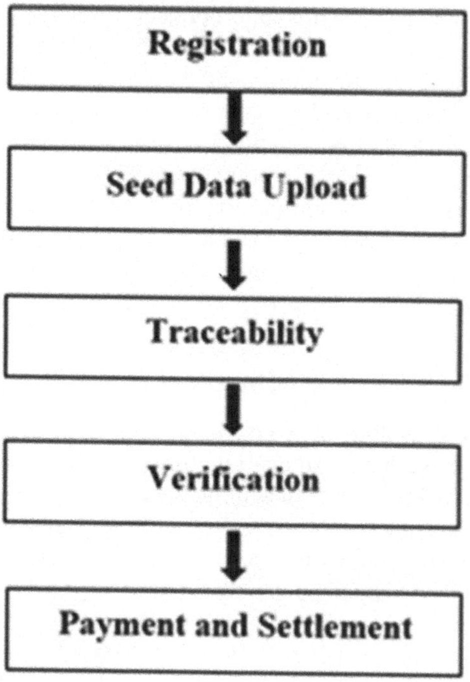

FIGURE 20.1 System workflow for blockchain-based seed management system.

The workflow begins with the registration of the seed in the system, followed by the uploading of data attributes associated with it onto distributed storage as shown in Figure 20.1. Tracing of all stages starts once the seed is entered into the blockchain network, with verification occurring at each stage as data points move through the system. Whenever a new purchase is made, transactions associated with payment settlement are created for each node in the system. Above all, steps and flow guides for the implementation of a blockchain-based seed management system that enables every stakeholder to perform the required actions without affecting their physical process in a virtual model as well, with more security and a seamless management process.

20.6 CASE STUDIES OR HYPOTHETICAL SCENARIOS

Following are simulated or real-time examples of the utilization of blockchain technology for seed management. In order to improve farmer confidence in purchasing new seed breeds, a project is currently being tested in India that uses blockchain to verify seed certification with traceability during supply chain movements [26,27]. In this method, a hash-based QR code is given to users with the seed batch, allowing them to track the origin and stages of distribution, along with verifying the certification associated with that particular seed.

Analysing the outcomes of this pilot project reveals several optimizations achieved through the use of blockchain for seed certification and tracing. First, it reduces expenditure and time by minimizing human resources involved in the process and eliminating delays caused by paperwork. Second, the immutable nature of blockchain ensures that only high-quality seeds, which meet the standards maintained by regulatory bodies, are passed to farmers. Finally, the accessibility of seed tracing data to farmers improves transparency in the process, enabling them to make informed decisions, which helps improve crop yields. By analysing this real-time application, it becomes clear that the integration of blockchain technology for seed management optimizes the farming process from various angles.

20.7 FUTURE TRENDS AND RESEARCH DIRECTIONS

Blockchain technology revolutionizes seed management systems by integrating with emerging technologies such as artificial intelligence (AI) and the IoT [28,29]. The following methods provide an overview of how these integrations can work. AI for predictive analytics in a blockchain network records large amounts of real-time data and stores data points for logging transactions that occur on the network. AI algorithms then analyse these large chunks of data, helping to develop predictive systems that assist farmers in making informed decisions on various processes such as resource allocation and seed breed selection, ultimately improving production. For example, an AI model trained on data generated from the blockchain network for seed management can identify patterns between production and seed utilization. This identified pattern can recommend the best-suited seeds for specific soil types, weather, or climate conditions.

Real-time data collection through IoT devices, such as soil sensors, moisture sensors, and trackers, helps in collecting data points on soil conditions, weather, and soil moisture levels from crop fields. The integration of blockchain ensures secure access to these data with proper access control mechanisms based on the role specified for each stakeholder, reducing the risk of data contamination shared between various users. The proposed blockchain network application in seed management has significant potential for regulatory compliance and quality assurance in seed management. Regulatory compliance within a blockchain network provides immutable records for seed production, certification, and distribution processes, ensuring that all local and global regulations are followed throughout the supply cycle.

This not only simplifies the audit process but also improves trust among stakeholders involved in the supply system. Quality assurance, on the other hand, is achieved through end-to-end tracing of seeds from production to purchase, managed by blockchain. Smart contracts for each node ensure automatic quality measures and guarantee that only certified, high-quality seeds reach the end farmers. However, despite these benefits, the implementation of blockchain in seed management faces major issues. One of the main challenges is the high expense of implementation, as creating, setting up, and maintaining blockchain is an expensive venture, especially for resource-intensive processes like seed management in the agricultural sector.

In order to overcome this issue, government bodies must encourage the implementation of subsidies. In addition, adoption barriers exist, as various stakeholders involved in seed supply management, such as small-scale suppliers and farmers, lack the technical skills and knowledge necessary to adapt to solutions like blockchain. This barrier can be overcome by offering mass education campaigns to provide skills and knowledge to these stakeholders.

20.8 CONCLUSION

The agricultural sector, burdened by concerns such as soil degradation, resource scarcity, and supply chain ineffectiveness, necessitates innovative solutions. Blockchain technology, with its intrinsic properties of security and immutability, offers a promising possibility to address these challenges. By allowing the creation of a devolved and auditable record of agricultural methods, from seed to shelf, blockchain can substantially enhance the sector's efficiency, consumer trust, and sustainability. The broader effects of blockchain for agriculture are far-reaching. It can encourage greater precision and accountability, reduce food fraud and forging, and enhance traceability. By optimizing supply chains, blockchain can lessen food waste and guarantee the timely delivery of fresh goods. Moreover, it can encourage farmers by granting them fair benefits and access to market statistics. In order to realize the capability of blockchain in agriculture entirely, a collective effort is required from various investors. Governments, industry leaders, policymakers, and technology providers must work collectively to develop strong blockchain solutions, verify regulatory frameworks, and encourage widespread acceptance. By adopting this transformative technology, we can make a more efficient, sustainable, and unbiased agricultural future.

REFERENCES

1. M. S. Farooq, Z. K. Ansari, A. Alvi, F. Rustam, I. D. L. T. Díez, J. L. V. Mazón, C. L. Rodríguez, and I. Ashraf, "Blockchain-based transparent and reliable framework for wheat crop supply chain," *PLoS One,* vol. 19, pp. 1–21, 2024, doi: 10.1371/journal.pone.0295036.
2. V. Puri, A. Kataria, and V. Sharma, "Artificial intelligence-powered decentralized framework for Internet of Things in Healthcare 4.0," *Trans. Emerg. Telecommun. Technol.,* vol. 35, no. 4, p. e4245, Apr. 2024, doi: 10.1002/ett.4245.
3. N. Singh, A. Sharma, Indraneel, S. Prabhud, and K.Chadaga, "IoT-based greenhouse technologies for enhanced crop production: A comprehensive study of monitoring, control, and communication techniques," *Syst. Sci. Contr. Eng.*, vol. 12, no. 1, Jan 2024, doi: 10.1080/21642583.2024.2306825.
4. R. Ahuja, S. Chugh, and R. Singh, "Seed chain: A secure and transparent blockchain-driven framework to revolutionize the seed supply chain," *Futur. Internet,* vol. 16, no. 4, pp. 1–20, 2024, doi: 10.3390/fi16040132.
5. R. K. Dewang, M. P. Yadav, S. Awasthi, O. Raj, A. Mewada, and K. L. Bawankule, "Data secure application: An application that allows developers to store user data securely using blockchain and IPFS," *Multimed. Tools Appl.*, vol. 83, no. 15, pp. 45491–45517, 2024.
6. N. Singh, H. P. Singh, A. Mishra, A. Khare, M. Swarnkar and S. K. Almas, "Blockchain cloud computing: Comparative study on DDoS, MITM and SQL injection attack," In *2024 IEEE International Conference on Big Data & Machine Learning (ICBDML)*, Bhopal, India, 2024, pp. 73–78, doi: 10.1109/ICBDML60909.2024.10577412

7. J. J. Lohith, S. Shreya, and J. L. Hamsa Priya, "Unlocking efficiency in agricultural supply chains: A secure and transparent approach through blockchain technology," *Int. Conf. Emerg. Technol. Comput. Sci. Interdiscip. Appl.*, vol. 2024, pp. 1–9, 2024, doi: 10.1109/ICETCS61022.2024.10544311.
8. U. Peddibhotla, S. C. Chandran, P. Kumar, and R. Kumar, "Enhanced supply chain management in Indian agriculture using ssi and blockchain leveraged by digital wallet," In *2024 16th International Conference on Communication Systems & Networks, COMSNETS 2024,* Bengaluru, India, pp. 427–429, 2024.
9. A. Nair, U. Peddibhotla, S. C. Chandran, and R. Kumar, "Convergence of IoT and blockchain ecosystem to ensure traceability and reliability in agricultural supply chain," In *2024 16th International Conference on Communication Systems & Networks, COMSNETS 2024,* Bengaluru, India, pp. 388–390, 2024.
10. K. G. Gulen and A. Karaagac, "Agricultural food supply chain with blockchain technology: A review on Turkey," *J. Glob. Strateg. Manag.,* vol. 2018, pp. 13–27, 2024, doi: 10.20460/jgsm.2023.314.
11. M. Al Hemairy, M. Abu Talib, A. Khalil, A. Zulfiqar, and T. Mohamed, "Blockchain-based framework and platform for validation, authentication & equivalency of academic certification and institution's accreditation: UAE case study and system performance," *Education and Information Technologies,* vol. 29, no. 14, pp. 1–30, 2024.
12. U. B. Nookaratnam, S. Srinivasu, P. L. N. R. Sagar, and M. Kalyani, "A study of block chain technology of farmers portal," *Int. J. Adv. Eng. Arch. Sci. Technol.*, vol. 2, no. 3, pp. 109–114, 2024.
13. T. K. Agrawal, V. Kumar, R. Pal, L. Wang, and Y. Chen, "Blockchain-based framework for supply chain traceability: A case example of textile and clothing industry," *Comput. Indus. Eng.*, vol. 154, pp. 107130, 2021.
14. S. N. Khan, F. Loukil, C. Ghedira-Guegan, E. Benkhelifa, and A. Bani-Hani, "Blockchain smart contracts: Applications, challenges, and future trends," *Peer-to-Peer Netw. Appl.*, vol. 14, pp. 2901–2925, 2021.
15. C. C. Ordoñez, G. R. Gonzales, and J. C. Corrales, "Blockchain and agricultural sustainability in South America: A systematic review," *Front. Sustain. Food Syst.*, vol. 8, p. 1347116, 2024.
16. H. R. Hasan, A. Musamih, K. Salah, R. Jayaraman, M. Omar, J. Arshad, and D. Boscovic, "Smart agriculture assurance: IoT and Blockchain for trusted sustainable produce," *Comp. Electr. Agr.*, vol. 224, p. 109184, 2024. https://doi.org/10.1016/j.compag.2024.109184
17. N. Mahalingam and P. Sharma, "An intelligent Blockchain technology for securing an IoT-based agriculture monitoring system," *Multimed. Tools Appl.*, vol. 83, no. 4, pp. 10297–10320, 2024.
18. R. S. Veena, G. Khambra, R. Sayal, N. Singh, et al., "An energy efficient IPv6 packet delivery outline for industrial IoT over G.9959 protocol based Wireless Sensor Network (WSN)," *SN Comput. Sci.*, vol. 5, p. 813, 2024, doi: 10.1007/s42979-024-03195-4.
19. M. A. Ansari and D. K. Singh, "Human detection techniques for real time surveillance: A comprehensive survey," *Multimed. Tools Appl.*, vol. 80, pp. 8759–8808, 2021, doi: 10.1007/s11042-020-10103-4
20. N. Singh, S. P. Sasirekha, A. Dhakne, B. V. Sai Thrinath, D. Ramya, and R. Thiagarajan, "IOT enabled hybrid model with learning ability for E-health care systems," *Meas.: Sens.*, vol. 24, pp.1–14, 2022, doi: 10.1016/j.measen.2022.100567.
21. I. Saputra, M. Wahyudi, R. K. Serli, A. Yoraeni, N. Ruhyana, and I. Sutedja, "Enhancing environmental data management through dynamic smart contract interaction on blockchain," In *2024 International Conference on Information Technology Research and Innovation (ICITRI),* Universitas Nusa Mandiri, Jakarta, Indonesia, 2024, pp. 341–346.

22. A. Padmavathi, G.C. Reddy, and G. Santhosh, "Blockchain-based credit system to increase the creditability through carbon offset level," In *2024 IEEE International Conference on Computing, Power and Communication Technologies (IC2PCT)*, Greater Noida, India, Feb. 2024, pp. 1181–1186.
23. S. Kably, N. Alaoui, M. Arioua, K. Chougdali, S. Khoulji, and M. D. Gómez-López, "Blockchain framework tailored for agricultural IoTs," In *International Work-Conference on the Interplay Between Natural and Artificial Computation,* Cham, Switzerland, pp. 461–473, 2024.
24. A. Pokale, H. Azath, V. Muneeswaran, and D. Saravanan, "Towards resilient agricultural supply chains: A framework for validating the impact of blockchain technology on sustainability," In *2024 IEEE International Students' Conference on Electrical, Electronics and Computer Science (SCEECS)*, Greater Noida, India, pp. 1–9, 2024.
25. X. Zinyan, S. Z. B. Omain, S. F. Khatib, and J. Lei, "Application of blockchain technology in agricultural product supply chain: Cite space bibliometric analysis based on WOS Database," *Int. J. Acad. Res Econ. Manag. Sci.*, vol. 13, no. 2, pp. 487–509, 2024.
26. A. H. C. Hung, "The curious case of Blockchain in rural China: Unravelling power, profit, and surveillance," *Big Data Soc.*, vol. 11, no. 2, p. 20539517241259674, 2024.
27. T. R. C. Konfo, A. B. P. Chabi, A. A. Gero, C. Lagnika, F. Avlessi, G. Biaou, and D. Sohounhloue, "Recent climate-smart innovations in agrifood to enhance producer incomes through sustainable solutions," *J. Agric. Food Res.*, vol. 15, p. 100985, 2024.
28. N. Singh and Y. Kumar, "Security and privacy requirements in 5G healthcare," In *Block Chain for 5G Healthcare Applications*, pp. 1–30, Nov. 2021. IET Publication, doi: 10.1049/PBHE035E_ch1.
29. R. Gao, S. Huang, and B. Li, "Green agri-food blockchain technology for investment decision-making under cost information constraints," *IAENG Int. J. Appl. Mathem.*, vol. 54, no. 1, pp. 82–91, 2024.

Index